FOOD SCIENCE AND TECHNOLOGY

MILK FAT

COMPOSITION, NUTRITIONAL VALUE AND HEALTH IMPLICATIONS

FOOD SCIENCE AND TECHNOLOGY

Additional books in this series can be found on Nova's website under the Series tab.

Additional e-books in this series can be found on Nova's website under the e-book tab.

NUTRITION AND DIET RESEARCH PROGRESS

Additional books in this series can be found on Nova's website under the Series tab.

Additional e-books in this series can be found on Nova's website under the e-book tab.

FOOD SCIENCE AND TECHNOLOGY

MILK FAT

COMPOSITION, NUTRITIONAL VALUE AND HEALTH IMPLICATIONS

LINDA B. RUDD
EDITOR

NOVINKA

New York

For permission to use material from this book please contact us:
Telephone 631-231-7269; Fax 631-231-8175
Web Site: http://www.novapublishers.com

LIBRARY OF CONGRESS CATALOGING-IN-PUBLICATION DATA

ISBN: 978-1-62618-353-7

Library of Congress Control Number: 2013933835

Published by Nova Science Publishers, Inc. † New York

CONTENTS

PREFACE

Milk is a natural, colloidal system in which fat is found in the interior of small, spherical droplets, formerly known as fat globules. The chemical composition of milk lipids in mammals varies and is affected by various factors such as feeding, breed and others. This book studies the composition, nutritional value and health implications of milk fat. Topics include milk composition disturbance and animal organism dysfunction caused by aflatoxins; size characterization of fat globules in dairy products by field flow fractionation; dairy system impacts on milk fat composition related to human health; and milk fat globules' microstructure.

Chapter 1 – Aflatoxins constitute toxic metabolites of the fungi *Aspergillus flavus* and *Aspergillus parasiticus* and are considered to be the mycotoxins, first detected and researched by the scientists. Aflatoxin consumption by animals or humans could lead to their organism function disorders, mainly depending on the duration of consumption and the quantity of the toxin. Such disorders, caused by aflatoxins and/or by other mycotoxins as well, could seriously affect animal production, the quality of the foods of animal origin and also the farmers' income. After aflatoxin B1 is consumed by lactating animals, it is metabolised to aflatoxin M1, which is excreted into the milk. As a consequence of aflatoxin influence on the general condition of animals, together with the excretion into the milk, the effects on milk quality, including changes in some milk components concentrations could be considered as possible. Unfortunately, the present literature regarding this suggestion is inadequate. The reports regarding aflatoxin effects on the general condition of animals are mainly related with basic parameters, whose changes could influence animal production. Such parameters are the body weight and the biochemical parameters of liver and kidneys. Reduction of body weight, decreased food consumption, increased concentrations of aspartate aminotransferase and probably of alanine aminotransferase, and extensive liver damage are reported by most authors. As for the milk fat content some researchers did not reveal any aflatoxin effects on fat values, while others report a significant reduction of milk fatty acids. These results were attributed to the inhibition or the decreased production of some enzymes, appropriate for fatty acids synthesis, such as synthetase. Regarding the milk protein content, no aflatoxin effects on it are reported. Most researchers make reason for a slight increase of total protein blood levels after aflatoxin consumption, as well as, for RNA polymerase suspended action, resulting in the inhibition of a lot of metabolic reactions, such as protein synthesis. Moreover, no significant changes were recorded concerning either the milk lactose content or the glucose blood concentration, after aflatoxin consumption. The likelihood of the possible aflatoxin effects on udder health, resulting in mastitis, or on somatic cell count and total microbial count changes cannot be supported by the up-today existing data. Additionally, these parameters could be affected by a variety of factors, requiring, for this reason, extensive research.

Chapter 2 – Milk is a natural, colloidal system in which fat is found in the interior of small, spherical droplets, formerly known as fat globules. The chemical composition of milk lipids in mammals varies and is affected by various factors such as feeding, breed and others. The main components of milk fat are triglycerides and they affect to a great extent the nutritional, organoleptic and physico-chemical properties of milk. Appearance and taste of milk food products are largely dependent on the particle size of the dispersed fat globules. Field Flow Fractionation (FFF) is a family of elution based separation techniques which is applied for the analysis of macromolecules, colloidal materials and proteins. The use of FFF methods for fractionation and physico-chemical characterization of food macromolecules has attracted an increasing interest during the last years. A wide range of Field flow fractionation sub-techniques have been employed for the study of dairy products, including particle size analysis of fat globules. In this paper, recent developments in the use of FFF methods for separation and size characterization of fat droplets are reviewed. Furthermore, the effects of milk processing on the particle size of fat globules is studied using various FFF techniques and the results are critically discussed. The main advantages and limitations of the techniques and instrumentation used for milk fat characterization are also presented.

Chapter 3 – Feed and feeding management differ significantly among dairy production systems with consequences on milk composition. Milk is increasingly produced in the temperate agricultural zones using year round total confinement of cows and to a lesser and shrinking extent, pasture systems which employ grazing for up to half the year. In this chapter the authors review the effects of the system of production on milk fat composition and implications for human health. On pasture, cows consume fresh forage with grain fed typically twice daily at milking compared to confined cows that are fed conserved forages, corn silage and grains in a total mixed ration. Fresh forage is high in α-linolenic acid. This polyunsaturated fatty acid as well as the products of its biohydrogenation vaccenic acid and rumenic acid, beneficial fatty acids, are incorporated into milk fat. Depressed ruminal pH combined with high ruminal passage rate in grazing cows can reduce the efficiency of biohydrogenation and increase transfer rate of dietary unsaturated fatty acids and their biohydrogenation intermediates to milk while depressing mammary synthesis of atherogenic saturated fatty acids, including palmitic acid. The authors' research shows that compared to confinement, the milk produced in a pasture system has a fatty acid profile characterized by a lower content of palmitic acid, a lower saturated to unsaturated fatty acids ratio, and a higher content of oleic acid, vaccenic acid and rumenic acid. Production of milk in pasture systems in temperate North America is a natural approach for improving the nutritional value of milk fat.

Chapter 4 – Milk fat globules are subject to several changes due to milk industrial processes and storage conditions. There is a potential industrial interest in monitoring changes in milk microstructure subject to various sanitation treatments and different storage temperatures. Most variations are related to the milk fat globules membrane (MFGM), whose main role is to keep isolated fat globules and protect them from lipolysis. Homogenization and heat treatments favour the deposit of milk proteins on the MFGM and their interactions, thus facilitating aggregation and fusion, and variation of size, shape and distribution of globules. Fat globules' structural properties can be studied by several techniques, such as X-ray diffraction and differential scanning calorimetry, scanning electron, fluorescence, and confocal microscopy, light scattering, SDS PAGE, and image analysis of milk micrographs. The authors review a number of techniques used to study the effects of industrial treatments on fat microstructure, aiming to detect measurable differences between strong and weak treatments.

In: Milk Fat
Editor: Linda B. Rudd

ISBN: 978-1-62618-353-7

Chapter 1

MILK COMPOSITION DISTURBANCE AND ANIMAL ORGANISM DYSFUNCTION CAUSED BY AFLATOXINS: A REVIEW

***Georgios D. Kourousekos*[1] *and Ekaterini K. Theodosiadou*[2]**
[1]School of Veterinary Medicine, Aristotle University of Thessaloniki, Thessaloniki, Greece
[2]Faculty of Veterinary Medicine, University of Thessaly, Karditsa, Greece

ABSTRACT

Aflatoxins constitute toxic metabolites of the fungi *Aspergillus flavus* and *Aspergillus parasiticus* and are considered to be the mycotoxins, first detected and researched by the scientists. Aflatoxin consumption by animals or humans could lead to their organism function disorders, mainly depending on the duration of consumption and the quantity of the toxin. Such disorders, caused by aflatoxins and/or by other mycotoxins as well, could seriously affect animal production, the quality of the foods of animal origin and also the farmers' income. After aflatoxin B1 is consumed by lactating animals, it is metabolised to aflatoxin M1, which is excreted into the milk. As a consequence of aflatoxin influence on the general condition of animals, together with the excretion into the milk, the effects on milk quality, including changes in some milk components concentrations could be considered as possible. Unfortunately, the present literature regarding this suggestion is inadequate. The reports regarding aflatoxin effects on the general condition of animals are mainly related with basic parameters, whose changes could influence animal production. Such parameters are the body weight and the biochemical parameters of liver and kidneys. Reduction of body weight, decreased food consumption, increased concentrations of aspartate aminotransferase and probably of alanine aminotransferase, and extensive liver damage are reported by most authors. As for the milk fat content some researchers did not reveal any aflatoxin effects on fat values, while others report a significant reduction of milk fatty acids. These results were attributed to the inhibition or the decreased production of some enzymes, appropriate for fatty acids synthesis, such as synthetase. Regarding the milk protein content, no aflatoxin effects on it are reported. Most researchers make reason for a slight increase of total protein blood levels after aflatoxin consumption, as well as, for RNA polymerase suspended action, resulting in the inhibition of a lot of metabolic reactions, such as protein synthesis. Moreover, no significant changes were recorded concerning either the milk lactose content or the glucose blood concentration, after aflatoxin consumption. The likelihood of the possible aflatoxin effects on udder health, resulting in mastitis, or on somatic cell count and total microbial count changes cannot be supported by the up-today existing data. Additionally, these parameters could be affected by a variety of factors, requiring, for this reason, extensive research.

Introduction - Basic Knowledge on Mycotoxins

Mycotoxins, as their name reveal, are toxins and specifically secondary metabolites produced by certain species of fungi. In the last years the presence of mycotoxins in feedstuffs and/or in foods and the research of their possible detrimental effects on animal's health and production, on the quality of foods of animal origin, as well as, on human health are systematically studied. The problem of mycotoxins is not a recent one. In the past decades mycotoxins were considered harmful just for the animals that consumed them and as a consequence for the economy of the farms. Also, in the past, the lack of conservation methods for human foods led to their immediate consumption and as a result there was not much time for fungi to grow and produce mycotoxins. Today, such conservation methods permit the long-lasting storage of foods and at the same time it is well realized that mycotoxins can be part of the food chain. Thus, the need for further research upon the mycotoxins and the problems they could cause remains of great importance.

The contamination of animals and/or human by the mycotoxins could occur through the contaminated feed or food consumption, but in recent years, some attention has been given to the contamination through inhalation of mycotoxins in indoor air, mainly those mycotoxins found in dust from agricultural environments [Hendry and Cole, 1993]. The consumption of mycotoxins with the feeds, mainly by the productive even by the companion animals [Leung et al., 2006], could lead to the disturbance of their organism normal function. The pathological conditions derived by mycotoxins are called mycotoxicoses [Whitlow and Hagler, 2005]. The symptoms of a mycotoxicosis depend on the type of mycotoxin; the amount and duration of the exposure; the age, the health and the sex of the exposed individual; and many poorly understood synergistic effects involving genetics, dietary status and interactions with other toxic insults [Bennett and Klich, 2003]. In general, mycotoxicoses could be acute, after the consumption of high mycotoxins concentrations that could result to the animal's death, or chronic, after lower concentrations of mycotoxins are consumed by an animal but for a longer period [Alexopoulos, 1997].

Table 1. The most important mycotoxins, the main fungi that produce them and the basic metabolites of each mycotoxin

Mycotoxins	Main fungi	Basic metabolites
Aflatoxins: B1, B2, G1 & G2	Aspergillus flavus Aspergillus parasiticus	M1 (metabolite of B1)
Zearalenone	Fusarium culmorum Fusarium graminearum	β-zearalenol α-zearalenol
Trichothecenes: Deoxynivalenol T2 toxin	Fusarium culmorum Fusarium graminearum Fusarium culmorum Fusarium graminearum Fusarium sporotrichioides	HT2 toxin
Ochratoxin A	Aspergillus ochraceus Penicillium verrucosum	
Fumonisins: B1, B2 & B3	Fusarium moniliforme Fusarium proliferatum	

Table 2. Usual effects occurring after the consumption of each mycotoxin by the animals

Mycotoxins	Usual effects
Aflatoxins	Reduction of feed consumption & metabolism Reduction of growth & production Immunosuppression against microbes & parasites
Zearalenone	False oestrus Enlarged mammary glands Pseudopregnancy Swollen external genital organs Reduction of fertility or infertility
Deoxynivalenol	Lower production Low feed convertibility Anorexia Vomiting
T2 toxin	Immunosuppression Reduction of growth Reduction of fertility
Ochratoxin A	Nephrotoxicity Carcinogenesis
Fumonisins	Lower production Pulmonary oedema (pigs) Hepatotoxicity (rats) Encephalomalacia (horses)

The most known mycotoxins are aflatoxins, zearalenone, trichothecenes, ochratoxin A and fumonisins. The researchers focus on the study of the mechanisms that mycotoxins act, on the way they are metabolized and on the possible effects on animal and human health. In Table 1, the most important mycotoxins, the main fungi that produce them and the basic metabolites of each mycotoxin are presented. Furthermore, Table 2 shows the usual effects occurring after the consumption of each mycotoxin by the animals.

Mycotoxins Production

The presence of fungi in feeds or foods is a prerequisite for mycotoxins production. Moreover, the occurrence of fungi spores in feeds or foods is crucial for fungi growth. The spores are easily transferred by air or water and they can invade the plants, during cultivation, or storage of feedstuffs. Appropriate environmental conditions favor fungi growth and proliferation. Thus, the fungi could occur either during cultivation and harvesting of plants or during feeds processing and storage. Additionally, after fungi's growth, their harmful metabolites could be produced, since the environmental conditions remain suitable. This means that the presence of fungi in feeds does not ensure mycotoxins occurrence and that the absence of mycotoxins from feeds does not imply the absence of fungi. Feeds or foods presenting low amount of fungi could not be considered as mycotoxin-free [Diekman and Green, 1992]. Mycotoxins are particularly stable during contaminated feed or food processing (e.g. thermal) [Scott, 1984; 1991]. Consequently, in order the feeds invasion by fungi to be avoided; special attention should be paid during feeds or food production, storage and processing. More specifically the potential risks of mycotoxins may be controlled by checking plant material for fungal contamination, by improving methods of cultivation, harvest and storage, by eliminating or diluting toxins from the contaminated food or feeds and by using adsorbents to reduce the bioavailability of toxins in the digestive tracts of animals [Yannikouris and Jouany, 2002]. Food contamination by mycotoxins has been reported worldwide, mostly in feeds that are susceptible to fungal growth, such as grains and cereals, leading to damage in health and economic losses in agriculture [Heidtmann-Bemvenuti et al., 2011]. The feedstuffs most often invaded by fungi are corn, wheat, oat, barley, cottonseed and soy-been. Fungi growth is influenced by the following factors:

- Humidity: fungi usually need high humidity but there are differences even among the same species
- Temperature: most fungi grow in environmental temperatures but some species could be developed under very low or very high temperatures
- Oxygen: fungi cannot grow without oxygen presence of at least 1-2%
- The sensitivity of certain hybrid plants [Diekman and Green, 1992].

The environmental conditions that could, theoretically, prevent fungi's growth on plants or in feedstuffs are presented below:

- Temperature < -2.2 °C
- Humidity $< 14\%$

- Relative humidity < 70%
- Oxygen < 0.5% [Diekman and Green, 1992].

AFLATOXINS

Aflatoxins are mycotoxins produced mainly by the fungi *Aspegillus flavus* and *Aspergillus parasiticus* [a(spergillus) + fla(vus) + toxin]. The most important aflatoxins are B1, B2, G1 and G2. Aflatoxins B1 and B2 present blue (B) fluorescence during liquid chromatography, while aflatoxins G1 and G2 present green (G). Aflatoxin B1 is the most potent and is detected more often in feedstuffs regarding the other aflatoxins [Diekman and Green, 1992].

The occurrence of aflatoxins in feedstuffs is not a rare phenomenon. In some cases they are detected in quite high concentrations. There are some reports presenting the death of animals by aflatoxins as "endemic". Osweiler and Trampel [1985] reported the death of 40 cows with clinical signs of aflatoxicoses. In that study the feedstuffs analysis revealed aflatoxins presence in high concentrations. Furthermore, Samarajeewa et al. [1975] reported several deaths of goats consuming aflatoxin-contaminated feedstuff. Feedstuff analysis showed the presence of 0.15 μg aflatoxin B1/g of feed and 0.23 μg aflatoxin G1/g of feed, while the goats had consumed that feedstuff for a time interval of at least 4 months. Similar results, but without so many animal deaths, are reported by Colvin et al. [1984] and Hall et al. [1989]. These researchers considered that heavy rains and high humidity observed prior the studies were responsible for fungi's growth in feedstuffs and aflatoxin production.

Today, most countries of the world have set maximum permissible levels regarding the presence of aflatoxins in feedstuffs or in human foods. For example, Food and Drug Administration (F.D.A.) in U.S.A. has set the above levels at 20 ppb either for the corn consumed by dairy cows or for human foods (F.D.A., 2000). In the European Union, the maximum limit of aflatoxin B1 in feedstuffs of dairy cows has been set at 5 ppb, while for human foods it ranges from 0.10 to 5 ppb [Commission Regulation EC, No 1881/2006 (E.C., 2006)].

EFFECTS OF AFLATOXINS ON ANIMAL'S GENERAL CONDITION

The consumption of aflatoxins by animals or human could lead to their organism dysfunction. Specifically, in animals, aflatoxins disturbances could be related to the reduction of production and to the reduction of products quality. Aflatoxins might cause death when consumed in high concentrations, while lower concentrations could have as a result chronic toxicosis even tumor development, mainly on the liver [Diekman and Green, 1992].

Ruminants are presented more resistant against aflatoxins compared with other animal species, mainly monogastric [Jouany et al., 2009]. The general clinical signs observed after chronic aflatoxin consumption or consumption in high concentrations by the animals are: reduction of feed consumption and metabolism, reduction of growth and production, immunosuppression against microbes and parasites, nasal excretion, dyspnea, cough, lethargy, as well as, hemorrhages in muscles [Clark et al., 1984; Diekman and Green, 1992; Jacobsen et al., 1993].

In most cases, the reports regarding the aflatoxins effects on animal's general condition are mainly based on the study of basic parameters, such as the body weight and some blood biochemical enzymes values, principally those of liver. The long-lasting change of such parameters could negatively impair animal production.

Regarding the body weight, most researchers observed reduction after aflatoxin administration to the animals. The decreased weight gain is often reported when aflatoxins are administered in young animals. Regarding ruminants, Clark et al. [1984] and Miller et al. [1984] recorded a reduction of the body weight in goats consuming aflatoxin B1. Fernandez et al. [1996] observed a significant decrease of lamb's body weight after aflatoxin administration. In the same study, the daily weight gain of lambs was also significantly reduced compared to that of controls. Moreover, liver weight of the animals receiving aflatoxin appeared particularly lower than that of controls. In another study of Fernandez et al. [2000] body weight reduction of lambs was observed after aflatoxin administration. Decrease of body weight was also recorded in cattle after aflatoxin consumption for 21 days [Neathery et al., 1980]. The cattle of that study presented decreased feed consumption and weight gain, while no significant observations were made on the liver of the animals. Contrastingly, in the research of Battacone et al. [2003], the body weight of ewes was increased after aflatoxin administration, while in goats

Kourousekos et al. [2012] found no significant effect of aflatoxin B1 administration in the body weight. Concerning other than ruminant's animal species, Nageswara et al. [1999] concluded a decrease in chicken's body weight after aflatoxin administration in feed for 15 days. The higher the aflatoxin administration was the greater the body weight decrease. Furthermore, when Doer and Ottinger [1980] administered aflatoxin in young quails for 3 weeks, they found a significant body weight reduction and a significant increase of liver weight. In the same study, there was also a significant decrease of testicles and ovaries weight of the animals receiving aflatoxin and that fact could influence their sexual maturation. Similar effects on body weight due to aflatoxin administration had been recorded by Dimitri et al. [1998] in rabbits.

As for the effects of aflatoxins on blood biochemical enzymes values, the researchers focus mainly on those enzymes related to the liver, such as: aspartate aminotransferase (AST), alanine aminotransferase (ALT), lactate dehydrogenase (LDH), alkaline phosphatase (ALP), etc. The results vary from study to study and from species to species. Regarding small ruminants, Clark et al. [1984] reported an increase in AST values and no effect on ALT and ALP values after aflatoxin B1 administration in male goats. Maryamma and Sivadas [1975] also mentioned an increase in AST values of goats receiving aflatoxin, while Battacone et al. [2003] observed increased ALT and ALP values and, contrastingly, no effect on AST after aflatoxin administration in ewes. Furthermore, Bingol et al. [2007] determined a decrease in ALT values of goats after aflatoxin administration. In the same study, a significant increase in LDH values of goats receiving aflatoxin was recorded. Also, in goats, AST and ALP presented significantly higher and lower, respectively, in the blood of animals receiving aflatoxin B1 in a dose dependent manner (50 or 100 μg/goat/day for 35 days) [Kourousekos et al. 2012]. In other animal species, the administration of aflatoxin in chickens led to a significant increase of AST and ALT values [Oguz et al. 2002]. The same conclusion was also revealed by the study of Madheswaran et al. [2004], in which quails received aflatoxin for 35 days. In the study of Aller et al. [1981] the administration of high aflatoxin concentration (2 mg/kg body weight/day for 5 days) in young horses resulted in the increase of most hepatic enzymes of the animals.

In all studies above, it seems that the effects of aflatoxins on body weight and/or on biochemical blood parameters (liver function) depend on the aflatoxin concentration administered, as well as, on the time interval of administration.

Table 3. Aflatoxin effects on body weight and blood serum biochemical enzymes values in various animal species

Animal species	Effects	Aflatoxin concentration administered	References
Goats	↓*, b.w.**, ↑*** AST[†]	0.1 mg/kg b.w./d[††] for 34 d	Clark et al. [1984]
	↓ b.w.	0.1 mg/kg b.w./d for 34 d	Miller et al. [1984]
	↑ LDH[£], ↓ ALT[‡]	32 - 470 ppb	Bingol et al. [2007]
	↑ AST, ↓ ALP[££] No effect in b.w.	50 or 100 µg/goat/d for 35 d	Kourousekos et al. [2012]
Sheep	↓ b.w.	2.5 ppm/d for 21 d	Fernandez et al. [1996]
	↓ b.w.	2 ppm/d for 37 d	Fernandez et al. [2000]
	↑ b.w., ↑ ALT & ALP	128 µg/d/animal for 14 d	Battacone et al. [2003]
Cattle	↓ b.w.	5 ppm/d for 21 d	Neathery et al. [1980]
	No effect in b.w. gain	0.5-2 µg/l of milk/d for 6 weeks	Van Dijk et al. [1984]
Chickens	↓ b.w.	0.5 or 1 mg/kg of feed for 15 d	Nageswara et al. [1999]
	↑ AST & ALT	50 or 100 ppb	Oguz et al. [2002]
Quails	↓ b.w.	5 or 10 µg/g of feed for 15 d	Doerr & Ottinger [1980]
	↑ AST & ALT	3 ppm/d for 35 d	Madheswaran et al. [2004]
Equine	↑ liver enzymes	2 mg/kg b.w./d for 5 d	Aller et al. [1981]
	Mainly hemorrhagic enteritis	2 mg/kg b.w., once	Bortell et al. [1983]
Rabbits	↓ b.w.	2 ppm/d/animal for 4 months	Dimitri et al. [1998]

* ↓: decrease.

** b.w.: body weight.

*** ↑: increase.

[†] AST: aspartate aminotransferase.

[††] d: days.

[£] LDH: lactate dehydrogenase.

[‡] ALT: alanine aminotransferase.

[££] ALP: alkaline phosphatase.

Some bibliographic details, regarding the effects of aflatoxins on body weight and blood hepatic enzymes values of different animal species, are presented in Table 3.

Furthermore, it is well stated that exogenous agents acting as endocrine disruptors, such as aflatoxins, may cause detrimental effects on the reproductive system. Endocrine disruptors may result in impaired reproduction, alterations in sexual differentiation, reduction of oocyte developmental competence, arrest of embryonic development, growth and development of certain cancers [Brevini et al., 2005; Sanderson, 2006; Crain et al., 2008]. Specifically, the effects of aflatoxins on the female reproductive system are related with alterations in: sexual maturation [Doer and Ottinger, 1980], growth and maturation of the follicles, concentrations of hormones [Ibeh and Saxena, 1997a; b; Kourousekos et al., 2008; 2012], gestation [Ibeh and Saxena, 1997a; b], and growth of foetus [Schmidt and Panciera, 1980; Arora et al., 1981]. Regarding the male reproductive system, most reports are related with the likely effect of aflatoxins on the size and weight of the genital organs [Sharlin et al., 1980], on spermatogenesis [Egbunike et al., 1980; Gopal et al., 1980], on the number and morphology of spermatozoa [Hafez et al., 1982; Agnes and Akbarsha, 2003], as well as, on the concentrations of hormones [Clarke and Ottinger, 1987].

Occurrence of Aflatoxins in Milk (Aflatoxin M1)

Aflatoxin M1 constitutes a hydroxylized metabolite of aflatoxin B1. Aflatoxin M1 is also called "milk toxin", since the highest concentrations are detected into the milk. Since milk is the base for other foods of animal origin, such as cheese, yoghurt or butter, aflatoxin M1 could be detected in these products as well.

Aflatoxin M1 is detected into the milk after lactating animals consume aflatoxin B1-contaminated feedstuffs. Aflatoxin M1 reaches the highest concentration in milk within 3 days, whereas it cannot be detected 4-5 days after aflatoxin-contaminated feedstuffs withdrawal [Diekman and Green, 1982]. In cows receiving 13 mg aflatoxin B1/day for 7 days, aflatoxin M1 was detected into the milk till the day 10, while the milk production and feed consumption by the animals were reduced [Applebaum et al., 1982].

Aflatoxin M1 has been detected into the milk of cows, ewes, goats, buffalos and camels [Galvano et al., 1996], and into that of women as well. In some cases, aflatoxin M1 has been detected in quite high concentrations into the milk of women, mainly from tropical or subtropical regions [El-Nezami et al., 1995; Saad et al., 1995; Galvano et al., 1996, Abdulrazzaq et al., 2003].

The proportion between aflatoxin B1 consumed and aflatoxin M1 excreted into the milk ranges from 58:1 to 75:1 (66:1, in average). This proportion varies from animal to animal, from day to day and among the milkings as well, and it depends on daily milk production by each animal [Harris and Staples, 1992]. In ewes, it seems that the above proportion is lower compared with that observed in cows and goats [Battacone et al., 2003].

Researchers around the world study the degree to which milk or milk products are contaminated by aflatoxin M1. Most reports regard pasteurized milk [Nakajima et al., 2004] compared with raw milk [Roussi et al., 2002; Hussain et al., 2010], because the first is mostly preferred by the consumers. In most studies, aflatoxin M1 occurs into the milk, but in concentrations lower than the maximum permissible level set by each country. For instance, in 207 out of 208 pasteurized milk samples in Japan, aflatoxin M1 was detected in quite lower concentrations than the country's maximum level [Nakajima et al., 2004]. Similar results were observed in Italy [Galvano et al., 2001], Serbia [Polovinski-Horvatovic et al., 2009], Kuwait [Srivastava et al., 2001] and Brazil [Garrido et al., 2003]. Higher aflatoxin M1 concentrations were recorded in pasteurized milk samples presenting higher fat content [Carvajal et al., 2003a]. Aflatoxin M1 detection below the maximum legal limits in ewes' raw milk did not present a serious human health hazard in Italy [Bognanno et al., 2006]. When 50 or 100 μg aflatoxin B1/animal/day, for 35 days, was administered in goats, aflatoxin M1 was detected into raw milk in concentrations exceeding the maximum European limit [Kourousekos et al., 2012]. Also, a high percentage (40%) of aflatoxin M1 concentrations exceeding the European Union's maximum tolerance limit was reported in raw milk obtained from dairy plants of Iran [Kamkar, 2005]. Furthermore, high aflatoxin M1 concentrations were determined in cow milk in regions with tropical or subtropical climate, such as in Northern Africa [Elgerbi et al., 2004]. Nemati et al. [2010] reported that, in 33% of the total milk samples, collected from local shops in a city of Iran and examined for aflatoxin M1 detection, aflatoxin M1 levels were higher than the maximum residue limit accepted by the European Union.

Either aflatoxins or their metabolites are really stable during milk processing. Aflatoxin M1 is not impaired by milk pasteurization and/or during cheese or yoghurt production [Govaris et al., 2001; 2002; Mendonca and Venancio, 2005]. According to the results of Harris and Staples [1992], aflatoxin M1 concentration in pasteurized milk was the same as that of raw milk after 17 days of storage in 39 ^{o}F (3,89 ^{o}C). Another aflatoxin B1 metabolite, aflatoxicol, also seems to be stable during pasteurization [Carvajal et al., 2003b].

During yoghurt production, the initial aflatoxin M1 concentration in milk might be reduced because of the action of Lactobacilli. It seems that aflatoxin M1 reduction during yoghurt production depends on pH. Govaris et al. [2002] revealed that aflatoxin M1 was decreased more during yoghurt production at pH 4.0 than at pH 4.6. Concerning cheese, aflatoxin M1 concentration was found higher at the initial stages of cheese production than after cheese maturation [Govaris et al., 2001]. But this action of Lactobacilli seems that cannot diminish aflatoxin M1. In Portugal, aflatoxin M1 was detected in market yoghurts [Martins and Martins, 2004], as well as, in Turkey [Aydemir Atasever et al., 2011]. Also, in Iran, aflatoxin M1 was detected above the national standard limits in yoghurt, white cheese and pasteurized milk samples, even in butter and ice cream samples [Fallah, 2010]. Aydemir Atasever et al. [2010] also detected aflatoxin M1 in butter samples in Turkey. Moreover, aflatoxin M1 was detected in infant milk products and liquid milk samples in India [Rastogi et al., 2004].

As far as concerning the milk yield, the administration of 0.11 mg aflatoxin B1/kg of body weight/day for 5 days, in 8 lactating cows revealed a decrease in milk production, attributed to a reduction in food consumption [McGrew et al., 1982]. Battacone et al. [2003] reported an increase of the milk production after aflatoxin B1 administration in sheep for 14 days, while in the same study the milk production of the control group decreased. No effect on milk yield was recorded by Kourousekos et al. [2012], when they administered aflatoxin B1 in goats for 35 days.

Regarding the relation between aflatoxins and mastitis, there is no evidence that the presence of aflatoxin in the udder could result in inflammation. In cows receiving aflatoxin and having udders contaminated with streptococcus and staphylococcus, reduction of milk production and increase of the microbial count of milk were observed, without any clinical signs of mastitis [Brown et al., 1981].

Like for aflatoxin B1, most countries of the world have set maximum desirable levels regarding the presence of aflatoxin M1 in milk. F.D.A. in U.S.A. has set the above level at 500 ppt (Food and Drug Administration, U.S.A., 2000), while the maximum permissible level for aflatoxin M1 in milk set by the European Union is 50 ppt [Commission Regulation EC, No 1881/2006 (E.C., 2006)].

EFFECTS OF AFLATOXINS ON MILK COMPOSITION

Basic Knowledge on Milk Composition

Fat, proteins and lactose constitute the basic milk chemical ingredients that are composed into the udder's cells. Proteins and lactose are formed into the udder by amino acids and glucose, respectively, derived from blood circulation. Fat is composed by fatty acids, a part of which is derived from blood circulation, while the rest of them are produced into the udder [Mantis, 1993; Smokovitis, 2007].

Specifically, milk fat contains either short-chain fatty acids (4-15 C, approximately 30% of the total fatty acids) or large-chain fatty acids (16 C or more, approximately 70% of the total fatty acids). It seems that the large-chain fatty acids occurring into the milk are derived from blood triglycerides, whereas the short-chain fatty acids are composed into the udder by acetate and β-hydroxybutyric acid. According to Dimick et al. [1970], a part of milk fatty acids with 16 C is also produced by β-hydroxybutyric acid. Gerson et al. [1968] found that some fatty acids with 16 C, but mainly those with 18 C, were derived from blood glycerides.

Milk composition is influenced by many factors, and as a result the content of each component varies from animal to animal and from day to day. Such factors are [Mantis, 1993]:

- The amount of milk production
- The stage of the lactation period
- The udder's health
- The climatic conditions
- The way of milking
- The nutrition.

For example, when the milk fat content depends on the amount of milk production, it is presented decreased if the milk quantity produced appears increased, while the milk fat content increases when the milk production is reduced [Zygogiannis and Katsaounis, 1992]. Furthermore, milk fat content increases at the end of the lactation period. For instance, goat milk fat content appears high during the first week of the lactation period (7.8%), while it is reduced after that week at the levels of 4.5-5.0% and it seems to be increased again after the 20th week of the lactation period [Zygogiannis and Katsaounis,

1986]. Regarding udder's health, Bansal et al. [2005] observed that during mastitis in cows, milk protein content and somatic cell count presented increased, while milk fat and lactose content decreased.

> Concerning milk lactose content, it was observed that it is influenced by the stage of the lactation period. High milk lactose content was recorded during the first 10 weeks postweaning, whereas it was reduced until the end of the lactation period of goats [Goulas, 1993]. According to other authors, milk lactose content remains stable during the whole lactation period [Mba et al., 1975; Boros, 1986].

The fact that aflatoxins could negatively impair the general condition of animals, in combination with aflatoxin M1 excretion into the milk, might lead to the consideration that aflatoxins could have some effects on milk quality, changing the content of some milk components. Unfortunately, the up-to-date existing data for such hypothesis are inadequate.

> Any possible change of some factors that impair milk composition could have as a result the increase or decrease of the content of the milk components, influencing, as a consequence, the quality and/or the trade value of milk. Taking into account: a) the often presence of aflatoxins in feedstuffs and milk, b) the effects of aflatoxins in the general condition of animals, c) the high milk nutritional value, and d) that the milk content depends on many different factors, some of the changes in milk content could be a direct or indirect result of aflatoxins action.

MILK FAT CONTENT

According to the present references, it seems that the milk fat content is impaired by aflatoxins through the reduced fatty acids synthesis, although some researchers did not report any effect of aflatoxins on the milk fat content (Table 4). McGrew et al. [1982] administered aflatoxin B1 or a combination of aflatoxin and phenobarbital in cows. They concluded that the total milk fat production did not change, although some years before Mertens and Wyatt [1976] observed the opposite. McGrew et al. attributed their results in the lower aflatoxin B1 dose administered and in the shorter time interval of aflatoxin administration. Nevertheless, in the same study, a significant decrease of the fatty acids, containing from 8 up to 14 C, was recorded. The authors considered as possible that aflatoxin could impair the fatty acids synthetase, and as a result the fatty acids production was reduced. They supported their hypothesis on the study of Donaldson et al. [1972], in which the administration of 10 mg aflatoxin B1/g feed for 3 weeks resulted in the inhibition in the action of the microsomal system of the fatty acids synthesis and of the fatty acids synthetase in the liver of chickens. These researchers proposed that aflatoxins reduce the formation of the enzymes that are responsible for fatty acids synthesis.

Many researchers tried to explain the mechanism through which aflatoxins could possibly influence the lipids metabolism. Many studies focus on possible negative effects of the aflatoxins in the availability of the substances that are important for the fatty acids synthesis and/or in the inhibition of the transformation of those substances to fatty acids. Black et al. [1970] observed a noticeable inhibition in the transformation of the above substances to phospholipids and cholesterol of the human skin because of the action of aflatoxin B1. The authors made the hypothesis that aflatoxin could influence a certain enzyme in lipids metabolism. Moreover, Tung et al. [1972] supported that aflatoxin could lead to a significant decrease of the blood serum triglycerides. Also, Kato et al. [1969] recorded an inhibition of cholesterol production due to aflatoxin and they supported that this inhibition was not a result of the general effect of aflatoxin on the hepatic function, since the rest enzymic systems of the liver remained normal.

The milk fat content appeared decreased, when goats received 50 or 100 μg aflatoxin B1/day/goat for 35 days, in a dose dependent manner [Kourousekos et al., 2012]. Battacone et al. [2003] administered aflatoxin B1 in ewes for 2 weeks and no significant differences of the milk fat content among the experimental groups was recorded. Smith et al. [1994] administered in goats 200 μg aflatoxin B1/kg feed/day for a shorter time interval (1 week) and they did not observe any significant differences regarding the milk fat content.

Milk Protein Content

The existing data, concerning the effects of aflatoxins on milk protein content, are limited. According to them, aflatoxins do not impair milk protein synthesis (Table 4). In the study of Battacone et al. [2003] the lowest milk protein content appeared in the milk of ewes that received the highest aflatoxin B1 dose. Smith et al. [1994] and Kourousekos et al. [2012] did not record any effect on milk protein content when goats were administered aflatoxin B1.

Generally, the most reports are related with the effects of aflatoxins on blood or tissue protein concentrations (Table 4). Clark et al. [1984] observed a slight increase in total proteins of the blood of goats receiving aflatoxin B1, but that increase was attributed to the dehydration of the animals due to the reduced water drinking. Also, Miller et al. [1984] reported a slight but not significant increase of the total blood proteins. However, it was found that aflatoxins could inhibit RNA polymerase action [Pong and Wogan, 1970] and bound RNA [Swenson et al., 1977] resulting in the inhibition of many metabolic functions, including that of protein or some enzymes synthesis [Wogan and Friedman, 1968; John and Miller, 1969; Kato et al., 1970].

Regarding other animal species, the administration of aflatoxin in laying hens for 3 weeks reduced the concentration of some vitamins of the group B and of some amino acids and as a result the inhibition of peptides formation was recorded [Voigt et al, 1980]. Also, Sidransky et al. [1977] and Bhat et al. [1982] concluded that aflatoxin inhibits the protein synthesis when they administered a high dose of the toxin in rats.

Milk Lactose Content

Regarding the milk lactose content, it does not seem that it could be influenced by aflatoxins (Table 4). Smith et al. [1994] and Kourousekos et al. [2012] did not observe significant differences in the milk lactose content of goats receiving aflatoxin B1. In the study of Battacone et al. [2003] the control group presented the lower lactose content, but this observation was not significant. Furthermore, it cannot be assumed if aflatoxins could impair the synthesis of lactose basic component, glucose. Applebaum and Marth [1983] administered in 7 cows 13 mg aflatoxin B1/day for 7 days and observed that only in one animal the blood serum glucose was reduced to a percentage of 9% during the experimentation period.

Milk hygiene - Udder's Health

Somatic cell count and total microbial count of the milk constitute parameters related directly with udder's health. These parameters can be influenced by many factors, more than those that could impair milk components, because somatic cell count and total microbial count depend also on a wide range of factors, such as the number of parturition or the stage of the lactation period [Boscos et al., 1996].

Consequently, the investigation of the possible effects of aflatoxins on these parameters is quite complicated. According to the existing data, aflatoxins could not be related with a possible mastitis occurrence. Battacone et al. [2003] and Kourousekos et al. [2012] did not observe significant differences in somatic cell count or in total microbial count among the experimental groups, while no clinical mastitis was recorded in ewes and in goats, respectively. In contrast, Brown et al. [1981] noticed an increase of the total microbial count in the milk of cows receiving 13 mg aflatoxin B1 *per os* for 7 days.

Table 4. Aflatoxin effects on milk's chemical components or on their production processes in animals

Milk components	Aflatoxin effects	References
Fat	Reduction of the total milk fat production	Mertens & Wyatt [1976]
	Reduction of the milk fatty acids, especially those containing 8-14 C	McGrew et al. [1982]
	Inhibition of the fatty acids synthesis	Donaldson et al. [1972]
	Decrease of the milk fat content	Kourousekos et al. [2012]
	No effect	Smith et al. [1994] Battacone et al. [2003]
Proteins	Increase of total proteins in blood	Clark et al. [1984]
	Inhibition of proteins or enzymes synthesis	Wogan & Friedman [1968] John & Miller [1969] Kato et al. [1970]
	Inhibition of peptides formation	Voigt et al. [1980]
	Inhibition of protein synthesis	Sidransky et al. [1977]
	No effect	Smith et al. [1994] Battacone et al. [2003] Kourousekos et al. [2012]
Lactose	Blood glucose reduction	Applebaum & Marth [1983]
	No effect	Smith et al. [1994] Battacone et al. [2003] Kourousekos et al. [2012]

Conclusion

Summarizing, it could be assumed that aflatoxins consumption by the animals, in high concentrations or for a long time interval, could affect their health. It seems that the body weight of the animals is reduced, while some biochemical parameters, mainly the liver function blood enzymes related, are increased, showing possibly hepatic dysfunction. Furthermore, it could not be concluded that aflatoxins impair the protein or lactose content of the milk or influence the udder's health. Contrastingly, a possible effect of aflatoxins on milk fat content could be supported, taking into consideration that aflatoxins could inhibit fatty acids synthesis and, thus, reduce their concentrations into the milk. Further research regarding the effects of aflatoxins on milk composition and/or on the mechanisms that they could affect milk components could be interestingly be useful, in order more specific conclusions to be made.

References

Abdulrazzaq, Y.M.; Osman, N.; Yousif, Z.M. & Al-Falahi, S. (2003). Aflatoxin M1 in breast-milk of U.A.E. women. *Ann Trop Paediatr,* 23, 173-179.

Agnes, V.F. & Akbarsha, M.A. (2003). Spermatotoxic effect of aflatoxin B1 in the albino mouse. *Food Chem Toxicol*, 41, 119-130.

Alexopoulos, K. (1997). Mycotoxicoses: The real problem in modern pig farming (in Greek). *Modern Pig Farming,* 4, 29-33.

Aller, W.W.Jr.; Edds, G.T. & Asquith, R.L. (1981). Effects of aflatoxins in young ponies. *Am J Vet Res,* 42, 2162-2164.

Applebaum, R.S.; Brackett, R.E.; Wiseman, D.W. & Marth, E.H. (1982). Responses of dairy cows to dietary aflatoxin: feed intake and yield, toxin content and quality of milk of cows treated with pure and impure aflatoxin. *J Dairy Sci,* 65, 1503-1508.

Applebaum, R.S. & Marth, E.H. (1983). Responses of dairy cows to dietary aflatoxin: Concentration of blood serum constituents and hormones associated with liver-kidney dysfunction and maintenance of lactation. *Appl Microbiol Biotechnol,* 18, 381-386.

Arora, R.G.; Frolen, H. & Nilsson, A. (1981). Interference of mycotoxins with prenatal development of the mouse. I. Influence of aflatoxin B1, ochratoxin A and zearalenone. *Acta Vet Scand,* 22, 524-534.

Aydemir Atasever, M.; Atasever, M.; Ozturan, K. & Urcar, S. (2010). Determination of aflatoxin M1 level in butter samples consumed in Erzurum, Turkey. *Kafkas Univ Vet Fak Derg,* 16, 159-162.

Aydemir Atasever, M.; Atasever, M. & Ozturan, K. (2011). Aflatoxin M1 levels in retail yoghurt and ayran in Erzurum in Turkey. *Turk J Vet Anim Sci,* 35, 59-62.

Bansal, B.K.; Hamann, J.; Grabowski, N.T. & Singh, K.B. (2005). Variation in the composition of selected milk fraction samples from healthy and mastitic quarters and its significance for mastitis diagnosis. *J Dairy Res,* 72, 144-152.

Battacone, G.; Nudda, A.; Cannas, A.; Cappio Borlino, A.; Bomboi, G. & Pulina, G. (2003). Excretion of aflatoxin M1 in milk of dairy ewes treated with different doses of aflatoxin B1. *J Dairy Sci,* 86, 2667-2675.

Bennett, J.W. & Klich, M. (2003). Mycotoxins. *Clin Microbiol Rev,* 16, 497-516.

Bhat, N.K.; Emeh, J.K.; Niranjan, B.G. & Avadhani, N.G. (1982). Inhibition of mitochondrial protein synthesis during early stages of aflatoxin B1 induced hepatocarcinogenesis. *Cancer Res,* 42, 1876-1880.

Bingol, N.T.; Tanritanir, P.; Dede, S. & Ceylan, E. (2007). Influence of aflatoxin present in forages and concentrated feedingstuffs on milk and some serum biochemical parameters in goats. *Bull Vet Inst Pulawy,* 51, 65-69.

Black, H.S.; Smith, J.D.; Austin, B.J. & Rauschkolb, E.W. (1970). Effect of aflatoxins B1 on the *in vitro* incorporation of ^{14}C-acetate into human skin lipids. *Experientia,* 26, 1292-1293.

Bognanno, M.; La Fauci, L.; Ritieni, A.; Tafuri, A.; Micari, P.; Di Renzo, L.; Ciappellano, S.; De Lorenzo, A.; Sarullo, V. & Galvano, F. (2006). Survey of the occurrence of Aflatoxin M1 in ovine milk by HPLC and its confirmation by MS. *Mol Nutr Food Res,* 50, 300-305.

Boros, V. (1986). Influence of the lactation period on variations in the levels of certain components of bulked goat's milk. *Int Dairy Fed Bull,* 202, 81-83.

Bortell, R.; Asquith, R.L.; Edds, G.T.; Simpson, C.F. & Aller, W.W. (1983). Acute experimentally induced aflatoxicosis in the weanling pony. *Am J Vet Res,* 44, 2110-2114.

Boscos, C.; Stefanakis, A.; Alexopoulos, C. & Samartzi, F. (1996). Prevalence of subclinical mastitis and influence of breed, parity, stage of lactation and mammary bacteriological status on Coulter Counter Counts and California Mastitis Test in the milk of Saanen and autochthonous Greek goats. *Small Rumin Res,* 21, 139-147.

Brevini, T.A.L.; Cillo, F.; Antonini, S. & Gandolfi, F. (2005). Effects of endocrine disruptors on the oocytes and embryos of farm animals. *Reprod Dom Anim,* 40, 291-299.

Brown, R.W.; Pier, A.C.; Richard, J.L. & Krogstad, R.E. (1981). Effects of dietary aflatoxins on existing bacterial intramammary infections of dairy cows. *Am J Vet Res,* 42, 927-933.

Carvajal, M.; Bolanos, A.; Rojo, F. & Mendez, I. (2003a). Aflatoxin M1 in pasteurized and ultrapasteurized milk with different fat content in Mexico. *J Food Prot,* 66, 1885-1892.

Carvajal, M.; Rojo, F.; Mendez, I. & Bolanos, A. (2003b). Aflatoxin B1 and its interconverting metabolite aflatoxicol in milk: the situation in Mexico. *Food Addit Contam,* 20, 1077-1086.

Clark, J.D.; Hatch, R.C.; Miller, D.M. & Jain, A.V. (1984). Caprine aflatoxicosis: experimental disease and clinical pathologic changes. *Am J Vet Res,* 45, 1132-1135.

Clarke, R.N. & Ottinger, M.A. (1987). The response of the anterior pituitary and testes to synthetic luteinizing hormone-releasing hormone (LHRH) and the effect of castration on pituitary responsiveness in the maturing chicken fed aflatoxin. *Biol Reprod,* 37, 556-563.

Colvin, B.M.; Harrison, L.R.; Gosser, H.S. & Hall, R.F. (1984). Aflatoxicosis in feeder cattle. *J Am Vet Med Assoc,* 184, 956-958.

Commission Regulation (EC), N° 1881 (2006). Setting maximum levels for certain contaminants in foodstuffs. *Official Journal of the European Union,* 364, 15.

Crain, A.; Janssen, S.; Edwards, T.; Heindel, J.; Ho, S.; Hunt, P.; Iguchi, T.; Juul, A.; McLachlan, J.; Schwartz, J.; Skakkebaek, N.; Soto, A.; Swan, S.; Walker, C.; Woodruff, T.; Woodruff, T.; Giudice, L. & Guillette, L.Jr. (2008). Female reproductive disorders: the roles of endocrine-disrupting compounds and developmental timing. *Fertil Steril*, 90, 911-940.

Diekman, M.A. & Green, M.L. (1992). Mycotoxins and reproduction in domestic livestock. *J Anim Sci,* 70, 1615-1627.

Dimick, P.S.; McCarthy, R.D. & Patton, S. (1970). Physiology of digestion and metabolism in the ruminants. In A.T. Phillipson, (Ed.), Proceedings of the Third International Symposium, *Cambridge August* 1969 (pp. 529-541). Cambridge, UK: Oriel Press, New Castle upon Tyne.

Dimitri, R.A.; Gabal, M.A. & Saleh, N. (1998). Effect of aflatoxin ingestion in feed on body weight gain and tissue residues in rabbits. *Mycoses,* 41, 87-91.

Doerr, J.A. & Ottinger, M.A. (1980). Delayed reproductive development resulting from aflatoxicosis in juvenile Japanese quail. *Poult Sci,* 59, 1995-2001.

Donaldson, W.E., Hsi-Tang Tung & Hamilton, P.B. (1972). Depression of fatty acid synthesis in chick liver (*Gallus domesticus*) by aflatoxin. *Comp Biochem Physiol,* 41, 843-847.

Egbunike, G.N.; Emerole, G.O.; Aire, T.A. & Ikegwuonu, F.I. (1980). Sperm production rates, sperm physiology and fertility in rats chronically treated with sublethal doses of aflatoxin B1. *Andrologia,* 12, 467-475.

Elgerbi, A.M.; Aidoo, K.E.; Candlish, A.A.G. & Tester, R.F. (2004). Occurrence of aflatoxins M1 in randomly selected North African milk and cheese samples. *Food Addit Contam,* 21, 592-597.

El-Nezami, H.S.; Nicoletti, G.; Neal, G.E.; Donohue, D.C. & Ahokas, J.T. (1995). Aflatoxin M1 in human breast milk samples from Victoria, Australia and Thailand. *Food Chem Toxicol,* 33, 173-179.

Fallah, A.A. (2010). Aflatoxin M1 contamination in dairy products marketed in Iran during winter and summer. *Food Control,* 21, 1478-1481.

Fernandez, A.; Hernandez, M.; Verde, M.T. & Sanz, M. (2000). Effect of aflatoxin on performance, haematology and clinical immunology in lambs. *Can J Vet Res,* 64, 53-58.

Fernandez, A.; Ramos, J.J.; Sanz, M.; Saez, T. & Fernandez de Luco, D. (1996). Alterations in the performance, haematology and clinical biochemistry of growing lambs fed with aflatoxin in the diet. *J Appl Toxicol,* 16, 85-91.

Food and Drug Administration, (F.D.A.), U.S.A. (2000) Guidance for Industry: Action Levels for Poisonous or Deleterious Substances in Human Food and Animal Feed. Available from: http://www.fda.gov/Food/GuidanceComplianceRegulatoryInformation/ GuidanceDocuments/ChemicalContaminantsandPesticides/ucm077969.htm.

Galvano, F.; Galofaro, V. & Galvano, G. (1996). Occurrence and stability of aflatoxim M1 in milk and milk products: a worldwide review. *J Food Prot,* 59, 1079-1090.

Galvano, F.; Galofaro, V.; Ritieni, A.; Bognanno, M.; De Angelis, A. & Galvano, G. (2001). Survey of the occurrence of aflatoxin M1 in dairy in dairy products marketed in Italy: second year of observation. *Food Addit Contam,* 18, 644-646.

Garrido, N.S.; Iha, M.H.; Santos Ortolani, M.R. & Duarte Favaro, R.M. (2003). Occurrence of aflatoxins M1 and M2 in milk commercialized in Ribeirao Preto-SP, Brazil. *Food Addit Contam,* 20, 70-73.

Gerson, T.; Shorland, F.B.; Wilson, G.F. & Reid, C.W.S. (1968). Origin of glyceride fatty acids in cow milk fat. *J Dairy Sci,* 51, 356-361.

Gopal, T.; Oehme, F.W.; Liao, T.F. & Chen, C.L. (1980). Effects of intratesticular aflatoxin B1 on rat testes and blood estrogens. *Toxicol Lett,* 5, 263-267.

Goulas, P. (1993). Breeding patterns and production of the indigenous goats (Capra prisca), kept under extensive and semi-intensive conditions in comparison with the Zaanen goats (in Greek). *PhD Thesis,* Faculty of Veterinary Medicine, Aristotle University of Thessaloniki, Greece.

Govaris, A.; Roussi, V.; Koidis, P.A. & Botsoglou, N.A. (2001). Distribution and stability of aflatoxin M1 during processing, ripening and storage of Telemes cheese. *Food Addit Contam,* 18, 437-443.

Govaris, A.; Roussi, V.; Koidis, P.A. & Botsoglou, N.A. (2002), Distribution and stability of aflatoxin M1 during production and storage of yoghurt. *Food Addit Contam,* 19, 1043-1050.

Hafez, A.H.; Megalla, S.E. & Mahmed, A.A. (1982). Aflatoxin and aflatoxicosis. III. Effect of dietary aflatoxin on the morphology of buffalo bull spermatozoa. *Mycopathologia,* 77, 141-144.

Hall, R.F.; Harrison, L.R. & Colvin, B.M. (1989). Aflatoxicosis in cattle pastured in a field of sweet corn. *J Am Vet Med Assoc,* 194, 938.

Harris, B.J. & Staples, C.R. (1992). The problems of mycotoxins in dairy cattle rations. University of Florida, Cooperative Extension Service, Institute of Food and Agricultural Sciences. Available from: www.edis.ifas.ufl.edu/BODY_DS152.

Heidtmann-Bemvenuti, R.; Mendes, G.L.; Scaglioni, P.T.; Badiale-Furlong, E. & Souza-Soares, L.A. (2011). Biochemistry and metabolism of mycotoxins: A review. *Afr J Food Sci,* 5, 861-869.

Hendry, K.M. & Cole, E.C. (1993). A review of mycotoxins in indoor air. *J Toxicol Environ Health,* 38, 183-198.

Hussain, I.; Anwar, J.; Asi, M.R.; Munawar, M.A. & Kashif, M. (2010). Aflatoxin M1 contamination in milk from five dairy species in Pakistan. *Food Control,* 21, 122-124.

Ibeh, I.N. & Saxena, D.K. (1997a). Aflatoxin B1 and reproduction. I. Reproductive performance in female rats. *Afr J Reprod Health,* 1, 79-84.

Ibeh, I.N. & Saxena, D.K. (1997b). Aflatoxin B1 and reproduction. II. Gametotoxicity in female rats. *Afr J Reprod Health,* 1, 85-89.

Jacobsen, B.J.; Bowen, K.L.; Shedly, R.A.; Diener, U.L.; Kemppainen, B.W. & Floyd, J. (1993). *Mycotoxins and Mycotoxicoses.* Alabama A&M and Auburn Universities. Available from: www.aces.edu/ department/ grain/ANR767.htm.

John, D.W. & Miller, L.L. (1969). Effect of aflatoxin B1 on net synthesis of albumin, fibrinogen and alpha1-acid glycoprotein by the isolated perfused rat liver. *Biochem Pharmacol,* 18, 1135-1146.

Jouany, J.P.; Yiannikouris, A. & Bertin, G. (2009). Risk assessment of mycotoxins in ruminants and ruminant products. *Options Méditerranées,* 85, 205-224.

Kamkar, A. (2005). A study on the occurrence of aflatoxin M1 in raw milk produced in Sarab city of Iran. *Food Control,* 16, 593-599.

Kato, R.; Onoda, K. & Omoki, Y. (1969). Effect of aflatoxins B1 on the incorporation of [^{14}C] acetate into cholesterol by rat liver. *Experentia,* 25, 1026.

Kato, R.; Takanaka, A.; Onoda, K. & Omori, Y. (1970). Different effect of aflatoxin on the induction of tryptophan oxygenase and of microsomal drug-hydroxylase system. *J Biochem,* 68, 589-592.

Kourousekos, G.D.; Theodosiadou, E.; Ververidis, H.; Belibasaki, S.; Boscos, C.; Saratsis, P. & Lymberopoulos, A.G. (2008) Effect of aflatoxin B1 administration on serum steroid hormones concentrations of goats. *Reprod Dom Anim,* 43 (S5), 102.

Kourousekos, G.D.; Theodosiadou, E.; Belibasaki, S.; Boscos, C. & Lymberopoulos, A.G. (2012). Effect of aflatoxin B1 on blood serum progesterone and oestradiol-17β concentrations of goats during anoestrus. *Reprod Dom Anim,* 47 (S5), 92.

Kourousekos, G.D.; Theodosiadou, E.; Belibasaki, S.; Deligiannis, K.; Koukoulas, Th.; Zoulfos, K. & Lymberopoulos, A.G. (2012). Effects of aflatoxin B1 administration on Greek indigenous goats' milk. *Int Dairy J,* 24, 123-129.

Leung, M.C.K.; Díaz-Llano, G. & Smith, T.K. (2006). Mycotoxins in Pet Food: A review on worldwide prevalence and preventative strategies. *J Agric Food Chem,* 54, 9623-9635.

Madheswaran, R.C.; Balachandran, B. & Murali, M. (2004). Influence of dietary culture material containing aflatoxin and T2 toxin on certain serum biochemical constituents in Japanese quail. *Mycopathologia,* 158, 337-341.

Mantis, A. (1993). *Hygiene and technology of the milk and its products* (2nd edition). Thessaloniki, Greece: Kyriakidis Bros.

Martins, M.L. & Martins, H.M. (2004). Aflatoxin M1 in yoghurts in Portugal. *Int J Food Microbiol,* 91, 315-317.

Maryamma, K.I. & Sivadas, C.G. (1975). Aflatoxicosis in goats (an experimental study). *Ind Vet J,* 52, 385-392.

Mba, A.I.; Boyo, B.S. & Oyenuga, V.A. (1975). Studies on the milk composition of West Africa dwarf, Red Sokoto and Saanen goats at different stages of lactation. I. Total solids, butterfat, solid-non-fat, protein, lactose and energy contents of milk. *J Dairy Res,* 42, 217-226.

McGrew, P.B.; Barnhart, H.M.; Mertens, D.R.; & Wyatt, R.D. (1982). Some effects of phenobarbital dosing of dairy cattle on aflatoxins M1 and fat in milk. *J Dairy Sci,* 65, 1227-1233.

Mendonca , C. & Venancio, A. (2005). Fate of aflatoxin M1 in cheese whey processing. *J Sci Food Agr,* 85, 2067-2070.

Mertens, D.R. & Wyatt, R.D. (1976). Acute aflatoxicosis in lactating dairy cows. *J Dairy Sci,* 60, 153-154.

Miller, D.M.; Clark, J.D.; Hatch, R.C. & Jain, A.V. (1984). Caprine aflatoxicoses: Serum electrophoresis and pathologic changes. *Am J Vet Res,* 45, 1136-1141.

Nageswara, A.R.; Reddy, V.R. & Srilatha, C. (1999). Effect of Livol Classic on the performance of commercial broilers fed aflatoxin. *Phytomedica,* 20, 65-70.

Nakajima, M.; Tabata, S.; Akiyama, H.; Itoh, Y.; Tanaka, T.; Sunagawa, H.; Tyonan, T.; Yoshizawa, T. & Kumagai, S. (2004). Occurrence of aflatoxin M1 in domestic milk in Japan during the winter season. *Food Addit Contam,* 21, 472-478.

Neathery, M.W.; Moos, W.H.; Wyatt, R.D.; Miller, W.J.; Gentry, R.P. & George, L.W. (1980). Effects of dietary aflatoxin on performance and zinc metabolism in dairy calves. *J Dairy Sci,* 63, 789-799.

Nemati, M.; Mehran, M.A.; Hamed, P.K. & Masoud, A. (2010). A survey on the occurrence of aflatoxin M1 in milk samples in Ardabil, Iran. *Food Control,* 21, 1022-1024.

Oguz, H.; Kurtoglu, F.; Kurtoglu, V. & Birdane, Y.O. (2002). Evaluation of biochemical characters of broiler chickens during dietary aflatoxin (50 and 100 ppb) and clinoptilolite exposure. *Res Vet Sci,* 73, 101-103.

Osweiler, G.D. & Trampel, D.W. (1985). Aflatoxicosis in feedlot cattle. *J Am Vet Med Assoc,* 187, 636-637.

Polovinski-Horvatovi, M.; Juri, V. & Glamoi, D. (2009). Two year study of incidence of aflatoxin M1 in milk in the region of Serbia. *Biotech Anim Husb,* 25, 713-718.

Pong, R.S. & Wogan, G.N. (1970). Time course and dose-response characteristics of aflatoxin B1 effects on rat liver RNA polymerase and ultrastructure. *Cancer Res,* 30, 294-304.

Rastogi, S.; Dwivedi, P.D.; Khanna, S.K. & Das, M. (2004). Detection of Aflatoxin M1 contamination in milk and infant milk products from Indian markets by ELISA. *Food Control,* 15, 287-290.

Roussi, V.; Govaris, A.; Varagouli, A. & Botsoglou, N.A. (2002). Occurrence of aflatoxin M1 in raw and market milk commercialized in Greece. *Food Addit Contam,* 19, 863-868.

Saad, A.M.; Abdelgadir, A.M. & Moss, M.O. (1995). Exposure of infants to aflatoxin M1 from mother's breast milk in Abu Dhabi, U.A.E. *Food Addit Contam,* 12, 255-261.

Samarajeewa, U.; Arseculeratne, S.N. & Tennekoon, G.E. (1975). Spontaneous and experimental aflatoxicosis in goats. *Res Vet Sci,* 19, 269-277.

Sanderson, T. (2006). The steroid hormone biosynthesis pathway as a target for endocrine-disrupting chemicals. *Toxicol Sci,* 94, 3-21.

Schmidt, R.E. & Panciera R.J. (1980). Effects of aflatoxins on pregnant hamsters and hamster fetuses. *J Comp Pathol,* 90, 339-347.

Scott, P.M. (1984). Effect of food processing on mycotoxins. J Food Prot, 47, 489-499.

Scott, P.M. (1991). Possibilities of reduction or elimination of mycotoxins present in cereal grains. In: J. Chelkowski (Ed.), *Cereal grains – mycotoxins, fungi and quality in drying and storage* (pp. 529-572). Amsterdam, The Netherlands: Elsevier.

Sharlin, J.S.; Howarth, B. Jr. & Wyatt, R.D. (1980). Effect of dietary aflatoxin on reproductive performance of mature White Leghorn males. *Poult Sci,* 59, 1311-1315.

Sidransky, H.; Verney, E.; Murty, C.N.; Sarma, D.S. & Reid, M. (1977). Effect of aflatoxin B1 on hepatic polyribosomes and protein synthesis in the rat. *Chem Biol Interact,* 18, 69-82.

Smith, E.E.; Phillips, T.D.; Ellis, J.A.; Harvey, R.B.; Kubena, L.F.; Thompson, J. & Newton, G. (1994). Dietary hydrated sodium calcium aluminosilicate reduction of aflatoxin M1 residue in dairy goat milk and effects on milk production and components. *J Anim Sci,* 72, 677-682.

Smokovitis A. (2007). *Physiology* (7th edition). Thessaloniki, Greece: Kyriakidis Bros.

Srivastava, V.P.; Bu-Abbas, A.; Allaa-Basuny; Al-Johar, W.; Al-Mufti, S. & Siddiqui, M.K.J. (2001). Aflatoxin M1 contamination in commercial samples of milk and dairy products in Kuwait. *Food Addit Contam* 18, 993-997.

Swenson, D.H.; Lin, J.-K.; Miller, E.C. & Miller, J.A. (1977). Aflatoxin B1-2, 3-oxide as a probable intermediate in the covalent binding of aflatoxins B1 and B2 to rat liver DNA and ribosomal RNA *in vivo*. *Cancer Res,* 37, 172-181.

Tung, H.; Donaldson, W.E. & Hamilton, P.B. (1972). Altered lipid transport during aflatoxicosis. *Toxicol Appl Pharmacol,* 22, 97-104.

Van Dijk, H.J.; O' Dell, G.D. & Bodine, A.B. (1984). Effects of aflatoxin M1 intake at physiologic levels on newborn dairy calves. *Am J Vet Res,* 45, 1994-1997.

Voigt, M.N.; Wyatt, R.D.; Ayres, J.C. & Koehler, P.E. (1980). Abnormal concentrations of B vitamins and amino acids in plasma, bile and liver of chicks with aflatoxicosis. *Appl Environ Microbiol,* 40, 870-875.

Whitlow, L.W. & Hagler, W.M. (2005). Mycotoxins: A review of dairy concerns. *Mid-South Ruminant Nutrition Conference,* pp. 47-58.

Wogan, G.N. & Friedman, M.A. (1968). Inhibition by aflatoxin B1 of hydrocortisone induction of rat liver tryptophan pyrrolase and tyrosine transaminase. *Arch Biochem Biophys,* 128, 509-516.

Yannikouris, A. & Jouany, J.P. (2002). Mycotoxins in feeds and their fate in animals: a review. *Anim Res,* 51, 81-99.

Zygogiannis, D. & Katsaounis, N. (1986). Milk yield and milk composition of indigenous goats (Capra prisca) in Greece. *Anim Prod,* 42, 365-374.

Zygogiannis, D. & Katsaounis, N. (1992). *Goat farming.* (1st edition). Thessaloniki, Greece: Kyriakidis Bros.

In: Milk Fat
Editor: Linda B. Rudd
ISBN: 978-1-62618-353-7

Chapter 2

SIZE CHARACTERIZATION OF FAT GLOBULES IN DAIRY PRODUCTS BY FIELD FLOW FRACTIONATION

Vassilios Raikos*
Department of Food Technology,
Technological Educational Institute of Kalamata, Kalamata, Greece

ABSTRACT

Milk is a natural, colloidal system in which fat is found in the interior of small, spherical droplets, formerly known as fat globules. The chemical composition of milk lipids in mammals varies and is affected by various factors such as feeding, breed and others. The main components of milk fat are triglycerides and they affect to a great extent the nutritional, organoleptic and physico-chemical properties of milk. Appearance and taste of milk food products are largely dependent on the particle size of the dispersed fat globules.

Field Flow Fractionation (FFF) is a family of elution based separation techniques which is applied for the analysis of macromolecules, colloidal materials and proteins. The use of FFF methods for fractionation and physico-chemical characterization of food macromolecules has attracted an increasing interest during the last years.

* Corresponding author: Vassilios Raikos, Technological Educational Institute of Kalamata, School of Agricultural Technology, Department of Food Technology, Kalamata, 24100, Greece. Tel/Fax: +30 27210 45237. E:mail:vas.raikos@gmail.com.

A wide range of Field flow fractionation sub-techniques have been employed for the study of dairy products, including particle size analysis of fat globules.In this paper, recent developments in the use of FFF methods for separation and size characterization of fat droplets are reviewed. Furthermore, the effects of milk processing on the particle size of fat globules is studied using various FFF techniques and the results are critically discussed. The main advantages and limitations of the techniques and instrumentation used for milk fat characterization are also presented.

1. Overview of Milk Fat: Structure, Composition and Functional Properties

Milk, because of its high nutritional value, occupies a central role in mammalian's diet. The role of milk in nature is not only to nourish but also to provide immunological protection for the young. Milk is a very complex food from a chemical point of view which contains varying amounts of water, proteins, lipids and carbohydrates and its composition can be affected by numerous factors such as breed variations, feed considerations, geographic locations, climate and others [Varnam & Sutherland, 1994].

Milk is very often described as a natural oil-in water emulsion, in which milk fat is dispersed in a continuous aqueous medium (Figure 1). The lipid fraction, which is mostly situated in small spherical oil droplets, is formerly known as fat globules and their size ranges from 0.1 to 15 μm in diameter [Heid & Keenan, 2005].These fat globules or droplets are covered by a thin biological membrane (milk fat globule membrane), which stabilizes the fat globule in the aqueous environment of milk [Keenan & Mather, 2002].

Fat is the most variable component in milk in terms of concentration and composition. Although the major source of variability in milk fat is feeding, genetic and environmental factors may account for differences in milk fat composition [Vanhoutte, 2002]. Triglycerides represent the most abundant class of lipids found in milk fat (95-98%). Triglycerides are composed of three fatty acids covalently bound to a glycerol molecule by ester bonds. According to Jensen [1992] over 400 different fatty acids have been identified in milk fat. As a result, this vast range and wide variety of fatty acids can combine in multiple ways to give a considerable number of possible molecular species of triglycerides synthesized in the mammary gland. According to Moore and Christie [1979] the distribution and arrangement of fatty acids on the triglyceride chain isnot random. This is indicated by enzyme mechanisms that

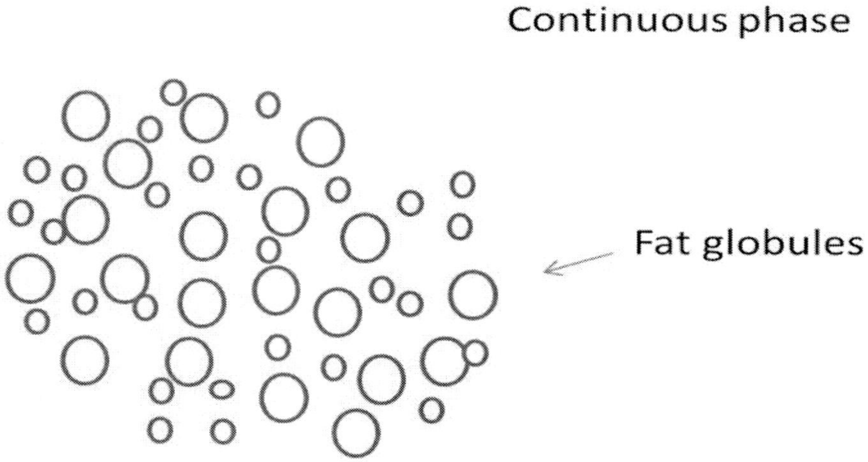

Figure 1. Milk fat globules dispersed in milk continuous phase (plasma).

display high selectivity with respect to the esterification of fatty acids at each position. Furthermore, although approximately 70% of the fatty acids in milk fat are saturated (medium and long chain) fatty acids such as palmitic acid and oleic acid, unsaturated and short chain fatty acids also exist [Sichien et al., 2009]. The non-triglyceride components present in milk fat, also known as minor componentsinclude mainly diglycerides, monoglycerides, phospholipids, sterols and free fatty acids (Table 1). Diglycerides, monoglycerides and free fatty acids may result from incomplete synthesis of triglycerides or due to lipolytic action, as indicated by their corresponding small amounts present in milk. The other minor components, such as phospholipids and cholesterol are situated in the milk fat globule membrane where they coexist with proteins and glycoproteins [Mezouari et al., 2009].

The primary biological role of milk fat is to be used from the mammalian newborn as a lipid source for the accumulation of body adipose tissue. Furthermore, milk fat provides a considerable amount of chemical energy to the young and also has a nutritional value as it supplies with essential fatty acids and fat soluble vitamins (A, D, E and K) the organism. Finally, the minor components of milk fat serve as emulsifiers and contribute to the uniform distribution of fat globules in the continuous phase.

Table 1. Composition of bovine milk lipids [Kurtz, 1978; Walstra et al., 2006]

Class of lipids	% Total milk lipids (w/w)
Triglycrides	95-98
Diglycerides	0.3-1.59
Monoglycerides	0.016-0.038
Keto acids glycerides (total)	0.85-1.28
Hydroxy acids glycerides (total)	0.60-0.78
Lactonogenic glycerides	0.06
Neutral glyceryl ethers	0.016-0.020
Neutral plasmalogens	0.04
Free fatty acids	0.1-0.44
Phospholipids (total)	0.80-1.00
Sphingolipids	0.06
Sterols	0.22-0.41
Squalene	0.03
Carotenoids	5-10 E^{-4}
Vitamin A	6-9 E^{-4}
Vitamin D	0.85-2.1 E^{-6}
Vitamin E	0.0024
Vitamin K	1 E^{-4}

2. Experimental Techniques for Characterization of Dairy Emulsions

An important part of processing of dairy products including milk is homogenization. During homogenization the fat droplets are disrupted into much smaller globules (below 1μm), by means of mechanical energy, in order increase the shelf-life of the product and consumer acceptability [Anderson et al., 1977]. This is because the size of the droplets in oil-in-water emulsions affects the stability, optical properties, rheology and sensory properties of the product [McClements, 2005]. Thus, the prerequisite for a high quality and cost effective end product in terms of appearance and taste is to reliably control the process of emulsion formation and stabilization (e.g. determination of droplet size) and accurately monitor changes caused by instability mechanismsduring storage (coalescence, flocculation, gravitational separation, phase inversion and ostwald ripening).

A wide range of techniques and sophisticated instruments are available today that can be employed to characterize emulsions. Among these, microscopy, particle size analysis and rheology are the most commonly applied for this purpose.

Microscopy

Various macroscopic techniques (optical, electron, atomic force) are routinely used to provide images of structurally complex foods such as emulsions. The size, shape and number of the droplets are either monitored directly by the eye lens of the microscope or most commonly by using a computer connected to the microscope with image processing software (Figure 2). Microscopy is a precise, reliable, relatively quick and inexpensive technique to study instability mechanisms in dairy emulsions such as flocculation. In particular, optical microscopy has been applied extensively in the past for this purpose with success [Dickinson, 2007]. The main limitation of this technique is the size of the droplets, which makes observation with an optical microscope a laborious task for samples with particle size below 1 μm. However, various techniques have been developed which aim to improve the overall image quality and thus narrow the limitations of this technique [Murphy, 2001]. Other limitations of microscopy for the characterization of dairy emulsions include sample preparation (e.g. dilution), which needs to be carried out carefully in order not to disturb the microstructure and the need to analyze multiple regions from the same sample to obtain statistically reliable data [McClements, 2007].

Particle Size Analysis

A variety of analytical instruments are available today to measure the "particle size distribution" of dairy emulsions, which refers to the concentration of droplets in different sizes [McClements, 2005]. Depending on the physical principles upon the relevant techniques operate, instrumentation varies considerably. Static light scattering (laser diffraction) for instance, is a very popular technique used to obtain particle size distribution data from food emulsions. The particular technique is based on the principle that the scattering pattern obtained when a laser beam passes through a dilute emulsion is related to the particle size distribution. The instrumentation used for laser diffraction is fully automated and provides rapid and precise measurements for particle size distributions, even when the size of droplets is below 1μm. However, the major limitation of this technique is that commercial instruments (e.g. Malvern Mastersizer) need to use a mathematical model to convert raw scattering data into the size distribution. This mathematical model, usually based on the Mie theory, makes several assumptions (e.g. assumes the droplets to be uniform spheres) which are not always valid and may lead to unreliable results [Dickinson, 1992]. Furthermore, the instrument requires additional input information (refractive index ratio, absorption coefficient) depending on the characteristics of the sample, to predict the particle size distribution. Other possible complication with this technique involves the sample preparation as emulsions need to be extensively diluted and sheared prior to analysis. This preparative step could affect the microstructure of the emulsion. Finally, sampling is very crucial in order to obtain a representative particle size distribution of the whole emulsion, especially for flocculated samples.

In addition to static light scattering, other techniques whichare commonly used to determine particle size distributions in food emulsions include dynamic light scattering, electrical pulse counting, ultrasonic spectrometry, NMR techniques and others [McClements, 2007].

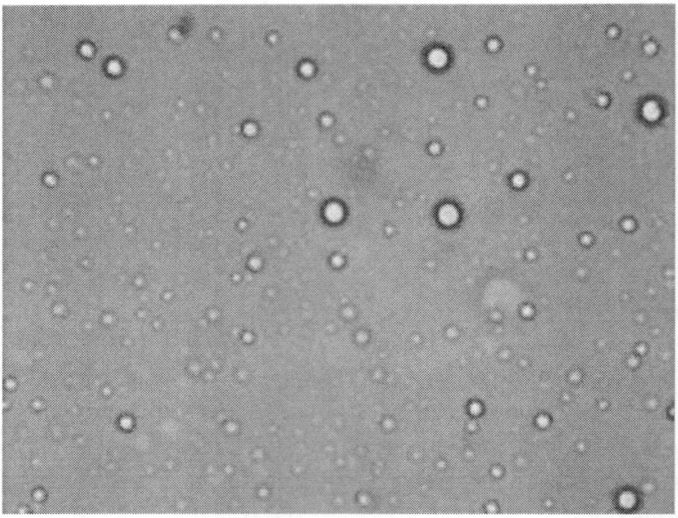

Figure 2. Milk fat droplets visualized with light microscope connected with a digital camera.

Rheology

Emulsion rheology is determined by physical factors such as droplet size distribution. Therefore, studying the rheology of a colloid system can provide useful knowledge in an attempt to understand its stability behavior [Dickinson & Stainsby, 1982]. Rheology is concerned with the relationship between the force per unit acting upon and within a material (stress) and the amount of deformation caused by the stress (strain). Most commonly in rheological studies, the strain varies and the resulting stress is monitored.

A wide range of instrumental methods are available today to determine the rheological properties of food emulsions, depending on the nature of the sample [Tadros, 2004]. Most commonly, shear devices can be used for analyzing fluid and viscoelastic devices whereas compression devices are preferred for testing viscoelastic, plastic or solid-like emulsions [Rao, 1999]. Rheological data has been used in the past to study the attractive forces between oil droplets in flocculated emulsions [Quemada & Berli, 2002]. Although emulsion rheology can provide fundamental information relative to the properties of a dairy product, instrumentation can be expensive, sophisticated and difficult to operate. In addition, experimental errors are quite common and special care and knowledge is required to interpret rheological data in order to eliminate effects of experimental artifacts [Macosko, 1994].

3. Field Flow Fractionation

Field flow fractionation (FFF) is a rapidly emerging family of separation techniques used for analysis of macromolecular, colloidal particles and proteins [Giddings, 1966; Giddings et al., 1980; Giddings, 1981; Giddings, 1993]. FFF is a high resolution technique offering experimental flexibility and redundancy and allows fractionation of a wide range of analytes suspended in both aqueous and organic solvent carriers. The main principle of FFF is based on a coupling between the parabolic velocity distribution of suspended particles through a ribbon-like channel and the application of a perpendicular, external field which compresses particles into layers against one wall of the channel (Figure 3). In a simplified way, the external field, which replaces the stationary phase of a typical chromatography technique, interacts with the sample components and is responsible for their separation by causing them to move perpendicular to the flow direction in the open channel. FFF instrumentation is combined with an appropriate detector for determination of

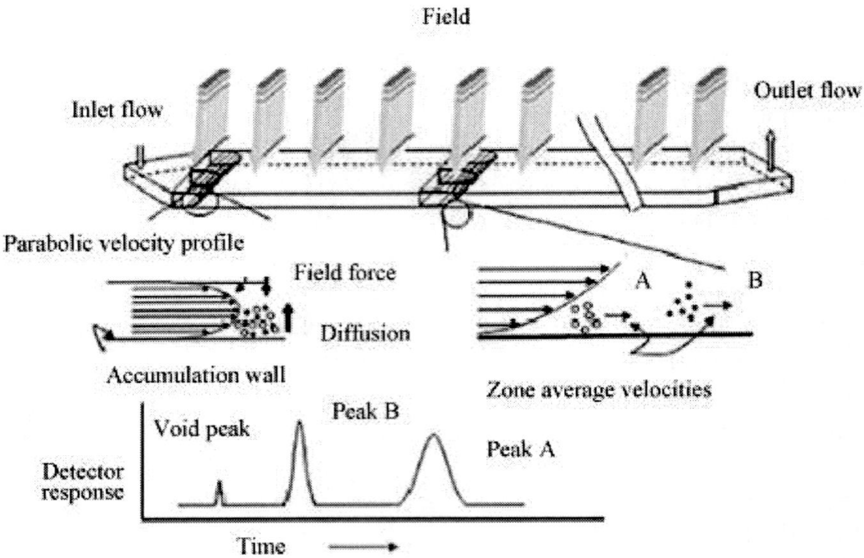

Figure 3. Schematic representation of an FFF channel, main principle of FFF separation and typical FFF fractogram. Reproduced from Messaud et al. (2009), with permission.

various molecular characteristics and usually include multi-angle light scattering (MALS), differential refractive index (dRI), differential viscometry, nuclear magnetic resonance (NMR), Fourier transform infrared (FTIR) or ultra-violet and visible (UV-VIS) [Messaud et al., 2009].

Various external fields can be used for particle separation and characterization in FFF and as a result various sub techniques have emerged. The external fields include sedimentation (sedimentation FFF), cross flow (flow FFF), temperature gradient (thermal FFF) and electrical field (field FFF) [Song et al., 1999]. Sedimentation FFF and flow FFF have been widely applied for the study of dairy products, and therefore their theory is further discussed in the following sections respectively (4 & 5).

FFF, which is a one-phase chromatographylike method, offers several advantages over other commercially available techniques for particle separation and characterization. FFF is an elution based technique and this allows the collection of well-defined fractions for further characterization. Furthermore, particles can be suspended in almost any medium of interest during experimentation to avoid problems associated with shrinking, swelling, dissolution or shape change [Giddings et al., 1983]. The amount of sample required for analysis is very small (<1mg) and the lack of stationary phase (open ribbon-like channel) minimizes the adsorption or shear degradation [Myers, 1997]. The intensity of the external fields applied is relatively mild and easy to control, thus allowing for analysis of complex, fragile samples [Williams & Lee, 2006]. One of the major limitations of FFF is the inability to analyze substances of low molecular weight and the time of analysis can be relatively long. Moreover, depending on the external field applied, additional requirements may limit the applications of FFF. In sedimentation FFF, for instance, the relative densities of the particles and the medium in which they are dispersed need to be known [Udabage et al., 2003].

4. Theory of Sedimentation FFF

In sedimentation FFF (SdFFF) separation is mediated by applying centrifugal forces on the particles suspended in a carrier liquid and colloidal particles are separated by differences in their buoyant mass [Giddings et al., 1983]. Suspended particles are compressed into layers against one wall of the channel, the mean thickness of which is determined by some field-coupled property (e.g. particle size) or a combination of such properties of the particles [Farmakis et al., 2000]. At the same time, a counteracting diffusive force acting on the particles drives the analyte back towards the center of the channel. Separation occurs because different particles reside in different flow velocity zones and these zones are slower compared to the average flow velocity of the carrier liquid passing through the channel [Messaud et al., 2009]. Particles concentrated in slower flow regions (closer to the wall) migrate more slowly compared to those with greater layer thickness which are found in regions near the channel center and thus, migrate faster (Figure 4). In the normal mode of sedimentation FFF, smaller particles travel faster than their larger counterparts, and the relationship between retention and particle size or mass can be predicted, considering that the force exerted on the sample particles by the applied centrifugal field is a calculable quantity. Sedimentation FFF yields experimental data in the form of a fractogram (plot of detector response of the eluting sample versus the retention time or volume) and particle size distributions are calculated on the basis of theoretical relationships [Koliadima & Karaiskakis, 1990].

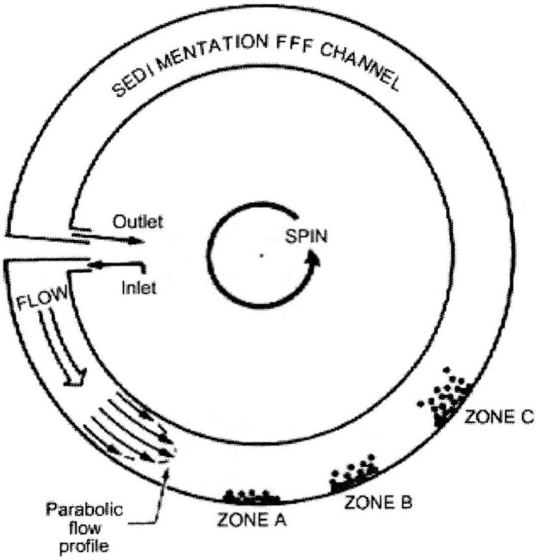

Figure 4. Schematic representation of an SdFFF channel. Particles are separated into zone layers according to their size (A→C, large → small). Reproduced from Raikos et al., (2009), with permission.

Provided that the exact geometry of the channel, the field strength, flow rate and the density difference between particle and carrier liquid are known, the diameter (d) of a spherical particle is given from the following equation [Dondi & Martin, 2000]:

$$d = \left(\frac{36kT}{\pi Gw\Delta\rho V^0}\right)^{1/3} Vr^{1/3} \qquad (1)$$

where k is the Boltzmann's constant, T is the absolute temperature, G is the angular acceleration (G= ω^2r, where ωis the angular velocity around radius r), w is the channel thickness, Δρ is the density difference between solute and solvent, V^0 is the column void volume and V_r is the retention time of particle under investigation.

5. Theory of Flow FFF

In flow FFF a cross-flow is applied as the external field perpendicular to the channel flow [Giddings, 1976]. In the case of symmetric flow FFF (FIFFF), cross-flow enters the upper block of the channel through the porous frit on the top wall and exits through an ultrafiltration membrane, overlaying a second porous frit at the bottom block (Figure 5). The separating force in FIFFF is the force exerted on sample components towards the bottom wall (accumulation wall) by the cross flow stream, where they are confined to a thin concentrated layer [Song et al., 1999]. Thecross-flow force is counteracted by diffusional transport at the accumulation wall and a steady state concentration distribution is obtained. As a result, following a relaxation period sample components are distributed according to the concentration average distance (l), from the accumulation wall. Small sample components are distributed further away from the accumulation wall compared to larger components, due to the higher diffusion coefficient of the smaller components [Nilsson, 2013]. Thus, as the longitudinal flow in the channel is laminar and the velocity is highest in the center of the channel, smaller particles move faster along the channel than larger components. The hydrodynamic diameter (d_h) is calculated from retention data using the following equation [Williams & Lee, 2006]:

$$d_h = (\frac{2kTV^0}{\pi n t^0 w^2 V_c}) t_r \qquad (2)$$

where k is the Boltzmann's constant, T is the absolute temperature, V^0 is the column void volume, n is the viscosity of the carrier liquid, t^0 is the void time, w is the channel thickness, V_c is the volumetric flow rate of the cross flow and t_r is the retention time of the sample component.

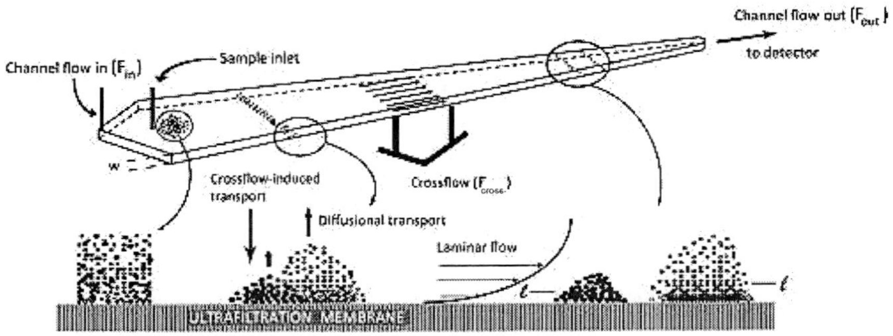

Figure 5. Schematic representation of FIFFF channel and principle of operation. Reprinted with permission from Nilsson (2013).

6. Size Characterization of Fat Droplets in Dairy Foods by Field Flow Fractionation

Milk fat is present in the form of globules with diameter up to 15μm. During processing, fat globules are reduced in size by means of mechanical action and their average diameter decreases to below 1μm. This processing step, also known as homogenization, aims to increase the stability of dairy foods and thus increase their shelf-life.

Field flow fractionation has been applied in the past to investigate the size of fat globules in commercially available products. Jussila et al. (1997) employed flow field-flow fractionation (FIFFF) to determine the particle size distribution of matriculate matter, such as fat globules, in a wide range of dairy products. The samples investigated in this study included homogenized dairy products such as full-fat milk, cream, ice cream, powdered milk, yoghurt and others. This study concluded that as expected in all dairy products the average size of fat globules was below 1μm, as revealed by the particle size distributions generated.

Table 2. Data on particle size of cereal and bovine milk samples generated with SdFFF. Reproduced with permission from Saeseaw et al. (2005)

	d_p (μm)	d_{mean} (μm)	Polydispersity (%CV)
Cereal milk			
Corn 1	0.40	0.50	68.0
Corn 2	0.38	0.47	64.9
Job's tear	0.33	0.38	53.9
Rice	0.31	0.38	65.8
Soy 1	0.32	0.38	57.9
Soy 2	0.32	0.42	66.7
Soy 3	0.33	0.44	68.2
Bovine milk			
Chocolate 1	0.42	0.50	59.0
Chocolate 2	0.44	0.53	52.8
Fresh full fat 1	0.46	0.51	52.0
Fresh full fat 2	0.46	0.52	47.1

These findings were in agreement with results published by researchers who employed other techniques, such as light scattering or Coulter counter, to investigate the fat globule distribution in homogenized milk [Zadow and Hardham, 1983; Anderson, 1991]. Sedimentation field flow fractionation (SdFFF) has also been applied in the past to study particle size distributions in dairy products. Saeseaw et al. [2005] used various types of milk, including cereal and bovine milk, in an attempt to investigate the applicability of SdFFF for milk fat globule characterization. The cereal milk samples included corn, job's tear, rice and soy milk, whereas chocolate and fresh full-fat milk were of bovine origin. In almost all cases, the homogenized samples exhibited monomodal distributions. As indicated by the results of this study, particle size distributions were in the range of 0.31-0.53μm, with the mean particle sizes of bovine milk being higher compared to the ones obtained for cereal milk (Table 2). The results of this study demonstrated the potential value of SdFFF for size characterization of complex food materials such as dairy products.

7. Field Flow Fractionation as a Tool for Studying the Impact of Processing on the Sizedistribution of Milk Fat Globules

Foodstuffs are commonly subjected to various processing steps in order to slow down the microbial spoilage and improve the physical properties of the end product. In the dairy industry, processing may include cooling, drying, separation, agitation, heating and homogenization [Lee & Sherbon, 2002]. During processing, the physico-chemical properties of the product may be altered. Thus, monitoring and controlling the effects of each processing step on the fat globule size distribution is a prerequisite for obtaining products with desired organoleptic characteristics, such as texture and mouthfeel.

Sedimentation field flow fractionation (SdFFF) was employed to investigate the effect of thermal processing on commercially available milk products [Raikos et al., 2009]. In this study, full-fat, semi-fat and chocolate milk of bovine origin was subjected to heat treatment in a temperature range between 50 and 125°C for 1h and the effect on particle size was studied. Figure 6 illustrates that SdFFF was applied successfully to determine the contribution of fat globules in milk, which corresponds to the peak eluting after 15 min, under the specified experimental conditions.

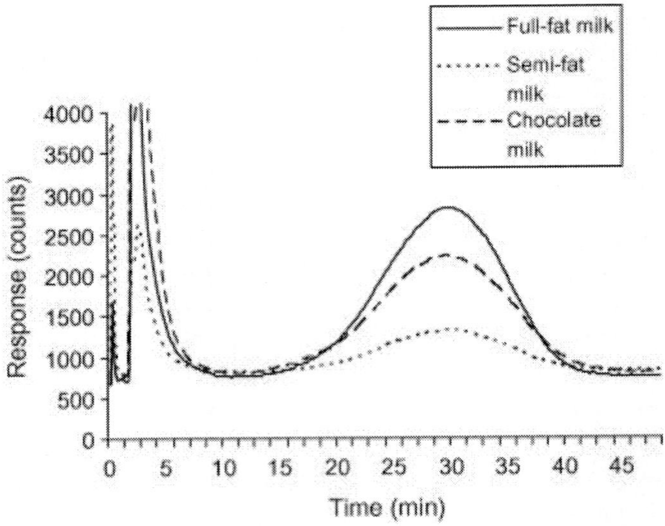

Figure 6. Representative SdFFF fractograms of full-fat (3.5%), semi-fat (1.0%) and chocolate milk (3.0%). The peaks correspond to the fat distribution of milk, as indicated by their relative heights. Reproduced from Raikos et al., (2009), with permission.

This study revealed that the mean particle diameter of fat droplets increases with increasing temperature, with chocolate milk being more affected from the processing step compared to full-fat and semi-fat milk (Figure 7). These findings are in agreement with research conducted by different methods and instrumentation and aimed to investigate the effect of heat treatment on the particle size of milk emulsions [Dickinson & Parkinson, 2004; Millqvist-Fureby et al., 2001]. Furthermore, SdFFF has also been employed to investigate the effect of pressure during the homogenization stage on the particle size distribution [Kenta et al., 2013]. In this study, model emulsions were formed using nonfat dried milk powder as protein source and were processed in a two-stage valve homogenizer in a pressure range from 20 to 60MPa. As revealed by SdFFF, increasing homogenization pressure, resulted in decreased particle size distribution of the oil-in-water emulsions, for the pressure range utilized in this study. According to this study, tripling the homogenizing pressure results in the formation of emulsions with oil droplet distributions reduced up to 55%. Sedimentation field flow fractionation was proven to be an adequate and reliable technique to monitor the changes in oil droplet size of milk protein stabilized emulsions as affected by homogenization pressure with high reproducibility (0.6-4.0%). Santana et al. (2011) used a different experimental approach (optical microscopy) and reported similar observations with respect to the effect of homogenization pressure on the particle size of oil-in-water emulsions. In conclusion, the results obtained by the different studies were comparable and particle size distributions of oil-in-water emulsions using SdFFF can be combined with images of the emulsion microstructures from optical microscopy in order to elucidate the effect of processing on the physico-chemical properties of complex colloidal food materials.

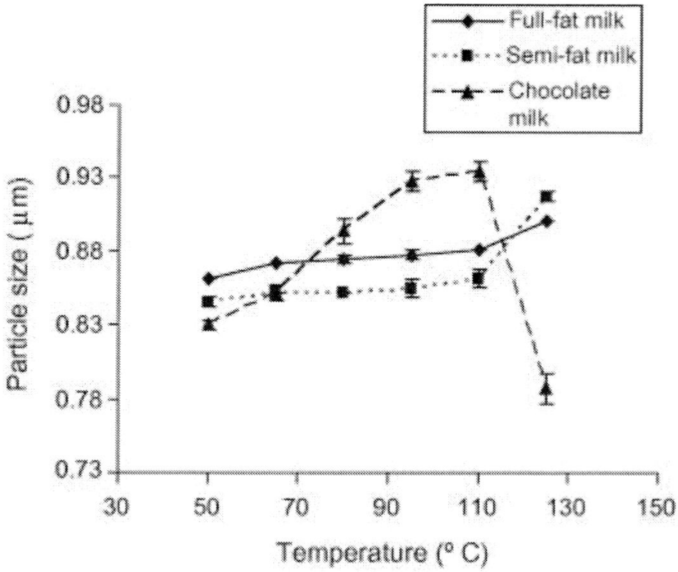

Figure 7. The effect of heating temperature on the mean particle diameter of full-fat, semi-fat and chocolate milk determined with SdFFF. Reproduced from Raikos et al., (2009), with permission.

Conclusion

Field flow fractionation is a well-established family of techniques suitable for colloid separation and characterization. The FFF techniques can be advantageous over conventional techniques used for milk fat globule size characterization, such as microscopy, because separated fractions can be collected and used for further analysis. This can be particularly useful for samples showing heterogeneity with respect to particle size and exhibit bimodal distribution. Laser light scattering is another very popular technique for determining particle size distributions in dairy products primarily because of the speed of analysis. Almost any type of particle can be measured using this technique, but the accuracy is dependent on knowing parameters such as the refractive index of the material and of the liquid. Furthermore, laser diffraction is not ideal for characterizing particles in the submicron range. On the other hand, FFF techniques exist which can be used for the non-destructive analysis of emulsion samples in the range of nm to μm and have inherent advantages for separation and size characterization of colloidal particles such as milk fat.

Thus, depending on the application, FFF techniques can be proven beneficial over other size characterization techniques, such as laser light scattering or microscopy, for overcoming limitations associated with high cost, inter-instrument reproducibility, resolution, submicron characterization and possible alteration of particles during sample preparation.In conclusion, this family of techniques can be a valuable tool for understanding the properties of colloidal systems and therefore its implementation in emulsion/dispersion technology laboratories is certainly justified.

REFERENCES

Anderson, M. (1991). Functional aspects of milk fat constituents. In K. K. Rajah & K. J. Burgess (Eds.), *Milk fat: production, technology and utilization* (pp. 9-17). Society of Dairy Technology, Huntington, Cambridgeshire, UK.

Anderson, M., Brooker, B. E., Cawston, T. E., & Cheeseman, G. C. (1977). Changes during storage in stability and composition of ultra-heat-treated aseptically-packed cream of 18% fat content. *Journal of Dairy Research*, 44, 111-124.

Dickinson, E. (1992). An introduction to food colloids. Oxford Science Publishers, Oxford, UK.

Dickinson, E. (1998). Proteins at interfaces and in emulsions: stability, rheology and interactions. *Journal of the Chemical Society, Faraday Transactions*, 94 (12), 1657-1669.

Dickinson, E. (2007). Colloidal systems containing droplets and bubbles. In D. J. McClements (Ed.), *Understanding and controlling the microstructure of complex foods* (pp. 153-184). Woodhead Publishing, Cambridge, UK.

Dickinson, E., & Parkisnson, E. L. (2004). Heat-induced aggregation of milk protein-stabilized emulsions: Sensitivity to processing and composition. *International Dairy Journal*, 14, 635-635.

Dickinson, E., & Stainsby, G. (1982). *Colloids in Food*. London: Applied Science.

Dondi, F., & Martin, M. (2000). Physicochemical measurements and distributionsfrom field flow fractionation. In M. E. Schimpf, J. C. Caldwell, & J. C. Giddings (Eds.), *Field flow fractionation handbook* (pp. 103–132). Wiley, New York.

Farmakis, L., Sakellaraki, J., Koliadima, A., Gavril, D., & Karaiskakis, G. (2000). Sizeanalysis of barley starch granules by sedimentation/steric field flowfractionation. *Starch/Stärk*e, 52, 275–282.

Giddings, J. C. (1966). A new separation concept based on a coupling of concentration and flow nonuniformities. *Journal of Separation Science*, 1, 1213-125.

Giddings, J. C. (1981). Field-flow fractionation: a versatile method for the characterization of macromolecules and particulate materials. *Analytical Chemistry*, 53, 1170A-1175A.

Giddings, J. C. (1993). Field-flow fractionation: analysis of macromolecular, colloidal, and particulate materials. *Science*, 260, 1456-1466.

Giddings, J. C., Karaiskakis, G., Caldwell, K. D., & Myers, M. N. (1983). Colloid characterization by sedimentation field-flow fractionation. I. Monodisperse populations. *Journal of Colloid and Interface Science*, 92, 66-80.

Giddings, J. C., Myers, M. N., Caldwell, K. D., & Fisher, S. R. (1980). Analysis of biological macromolecules and particles by field-flow fractionation. In D. Glick (Ed.), *Methods of biochemical analysis* (pp. 79-136). John Wiley & Sons, New York.

Giddings, J. C., Yang, F. J., & Myers, M. N. (1976). Theoretical and experimental characterization of flow field-flow fractionation. Analytical Chemistry, 48, 1126-1132.

Heid, H. W., & Keenan, T. W. (2005). Intracellular origin and secretion of milk fat globules. *European Journal of Cell Biology*, 84(2-3), 245-258.

Jensen, R. G. (1992). Fatty acids in milk and dairy products. In C.K. Chow (Ed.), *Fatty acids in foods and their health implications* (pp. 109-124). Marcel Dekker, New York.

Jussila, M. A., Yohannes, G., & Riekkola, M. L. (1997). Flow field-flow fractionation in the study of dairy products. *Journal of Microcolumn Separations*, 9 (8), 601-609.

Keenan, T. W., & Mather, I. H. (2002). Milk fat globule membrane. In H. Roginski, J. Fuquay, & P. F. Fox (Eds.), *Encyclopedia of dairy science* (pp. 1568-1576). Academic Press, London.

Kenta, S., Raikos, V., Kapolos, J., Koliadima, A., & Karaiskakis, G. (2013). Sedimentation field-flow fractionation as a tool for the study of milk protein-stabilized model oil-in-water emulsions: Effect of protein concentration and homogenization pressure. *Journal of Liquid Chromatography and Related Technologies*, Accepted manuscript (DOI: 10.1080/10826076.2011.653853).

Koliadima, A., & Karaiskakis, G. (1990). Potential-barrier field-flow fractionation, a versatile new separation method. *Journal of Chromatography*, 517, 345-359.

Kurtz, F. E. (1978). *The lipids of milk: Composition and properties*. In B. H. Webb, A. H. Johnson, & J. A. Alford (Eds.), Fundamentals of Dairy Chemistry (pp. 125-219), The Avi Publishing Company, Westport.

Lee, S. J., & Sherbon, J. W. (2002). Chemical changes in bovine milk fat globule membrane caused by heat treatment and homogenization of whole milk. *Journal of Dairy Research*, 69 (4), 555-567.

Macosko, C. W. (1994). *Rheology: Principles, measurements and applications*. VCH Publishers, New York.

McClements, D. J. (2005). *Food emulsions: Principles, practice and techniques*. Boca Raton, CRC Press.

McClements, D. J. (2007). Critical reviews of techniques and methodologies for characterization of emulsion stability. *Critical Reviews in Food Science and Nutrition*, 47, 611-649.

Messaud, F. A., Sanderson, R. D., Runyon, J. R., Otte, T., Pasch, H., & Williams, S. K. R. (2009). An overview on field-flow fractionation techniques and their applications in the separation and characterization of polymers. *Progress in Polymer Science*, 34, 351-368.

Mezouari, S., & Pouliot, Y. (2009). Lipids from the milk fat globule membrane as a health ingredient: Composition, properties and technological aspects. In M. Corredig (Ed.), *Dairy-derived ingredients* (pp. 344-367). Woodhead Publishing Limited and CRC Press LLC.

Millqvist-Fureby, A., Elofsson, U., & Bergenstähl, B. (2001). Surface composition of spray-dried milk protein-stabilized emulsions in relation to pre-heat treatment of proteins. *Colloids and Surfaces B: Biointerfaces*, 21, 47-58.

Moore, J. H., & Christie, W. W. (1979). Lipid metabolism in the mammary gland of ruminant animals. *Progress in Lipid Research*, 17, 347-395.

Murphy, D. B. (2001). *Fundamentals of light microscopy and electronic imaging*. John Wiley and Sons, New York.

Myers, M.N. (1997). Overview of field-flow fractionation. *Journal of Microcolumn Separations*, 9, 151-162.

Nilsson, L. (2013). Separation and characterization of food macromolecules using field-flow fractionation: A review. *Food Hydrocolloids*, 30, 1-11.

Quemada, D., & Berli, C. (2002). Energy of interaction in colloids and its implications in rheological modeling. *Advances in Colloid and Interface Science*, 98, 51-55.

Raikos, V., Kapolos, J., Farmakis, L., Koliadima, A., & Karaiskakis, G. (2009). The use of sedimentation field-flow fractionation in the size characterizationof bovine milk fat globules as affected by heat treatment. *Food Research International*, 42, 659-665.

Rao, M. A. (1999). *Rheology of fluids and semisolid foods: Principles and applications*. Springer, New York.

Saeseaw, S., Shiowatana, J., & Siripinyanond, A. (2005). Sedimentation field-flowfractionation: Size characterization of food materials. *Food Research International*, 38, 777–786.

Santana, R. C., Sato, A. C. K., & Cunha, R. L. (2012). Emulsions stabilized by heat-treated collagen fibers. *Food Hydrocolloids*, 26 (1), 73-81.

Sichien, M., Thienpont, N., Fredrick, E., Trung Le, T., van Camp, J., & Dewettinck, K. (2009). Processing means for milk fat fractionation and production of functional compounds. In M. Corredig (Ed.), *Dairy-derived ingredients* (pp. 68-102). Woodhead Publishing Limited and CRC Press LLC.

Song, J. H., Kim, W. S., Park, Y. H., Yu, E. K., & Lee, D. W. (1999). Retention characteristics of various proteins in flow field-flow fractionation: effects of pH, ionic strength, and denaturation. *Bulletin of the Korean Chemical Society*, 20(10), 1159-1164.

Tadros, T. (2004). Application of rheology for assessment and prediction of the long-term physical stability of emulsions. *Advances in Colloid and Interface Science*, 108-09, 227-258.

Udabage, P., McKinnon, I. R., & Augustin, M. A. (2003). The use of sedimentation fieldflow fractionation and photon correlation spectroscopy in the characterizationof casein micelles. *Journal of Dairy Research*, 70, 453–459.

Vanhoutte, B. (2002). *Milk fat crystallization: fractionation and texturisation.* PhD thesis, Ghent University, Belgium.

Varnam, A. H., & Sutherland, J. P. (1994). Milk and milk products: technology, chemistry and microbiology. Chapman & Hall, New York.

Walstra, P., Wouters, J. T. M., & Geurts, T. J. (2006). *Dairy Technology: Principles of Milk Properties and Processes*, CRC Press, New York.

Williams, S. K. R., & Lee, D. (2006). Field-flow fractionation of proteins, polysaccharides, synthetic polymers, and supramolecular assemblies. *Journal of Separation Science*, 29, 1720-1732.

Zadow, J. G., & Hardham, J. F. (1983). The stability of goat's milk to UHT processing. *Australian Journal of Dairy Technology*, 38, 20-23.

In: Milk Fat
Editor: Linda B. Rudd
ISBN: 978-1-62618-353-7

Chapter 3

Dairy System Impacts on Milk Fat Composition Related to Human Health

Payam Vahmani,[1] Alan Fredeen[2,*] and Kathleen Glover[3]

[1]Department of Biology, Dalhousie University, Halifax, Nova Scotia, Canada
[2]Department of Plant and Animal Sciences, Dalhousie University, Truro, Nova Scotia, Canada
[3]Azar Agriculture, Truro, Nova Scotia, Canada

Abstract

Feed and feeding management differ significantly among dairy production systems with consequences on milk composition. Milk is increasingly produced in the temperate agricultural zones using year round total confinement of cows and to a lesser and shrinking extent, pasture systems which employ grazing for up to half the year. In this chapter we review the effects of the system of production on milk fat composition and implications for human health. On pasture, cows consume fresh forage with grain fed typically twice daily at milking compared to confined cows that are fed conserved forages, corn silage

* Corresponding author: Alan Fredeen. Department of Plant and Animal Sciences, Dalhousie University, Truro, Nova Scotia, Canada B2N5E3. E-mail: alan.fredeen@dal.ca.

and grains in a total mixed ration. Fresh forage is high in α-linolenic acid. This polyunsaturated fatty acid as well as the products of its biohydrogenation vaccenic acid and rumenic acid, beneficial fatty acids, are incorporated into milk fat. Depressed ruminal pH combined with high ruminal passage rate in grazing cows can reduce the efficiency of biohydrogenation and increase transfer rate of dietary unsaturated fatty acids and their biohydrogenation intermediates to milk while depressing mammary synthesis of atherogenic saturated fatty acids, including palmitic acid. Our research shows that compared to confinement, the milk produced in a pasture system has a fatty acid profile characterized by a lower content of palmitic acid, a lower saturated to unsaturated fatty acids ratio, and a higher content of oleic acid, vaccenic acid and rumenic acid. Production of milk in pasture systems in temperate North America is a natural approach for improving the nutritional value of milk fat.

INTRODUCTION

Milk is a valuable component of human diets worldwide. One of the major components of milk is its fat which consists mostly of triacylglycerol molecules (TAGs) containing over 400 types of fatty acid (FA), the majority of which are derived directly and indirectly from the diet and from mammary synthesis [20, 31]. These nutrients, above all others in milk that can be modified by feeding management of the cow have the greatest impact on determining milk's healthfulness in the human diet.

Due to differences in feeding management, the FA composition of milk fat varies among dairy production systems which range from total confinement, in which a total mixed ration (TMR) is fed, to those that are pasture-based. The predominant system (confinement) has developed rapidly over the last few decades without concern for the impact of this change on the nutritional value of milk beyond its fat content. Some FAs in milk fat are considered beneficial to consumers including vaccenic acid (VA; *trans*-11 18:1), rumenic acid (RA; *cis*-9, *trans*-11 18:2), α-linolenic acid (ALA; 18:3 n-3) and oleic acid (OA; *cis*-9 18:1). Others are detrimental including the FAs 12:0-16:0 which can be synthesized in the mammary [31, 32]. A closer look at how the feeding management system affects the healthfulness of milk is needed.

Over the last 60 years, the source of forage in dairy systems in temperate North America has shifted from that of grazed pasture to silage. One main driver of this shift has been the favorable milk price to feed cost ratio, particularly between 1940 and 1980 when the focus of the dairy industry switched from production per unit of land to production per cow [1]. Pasture-based systems remain important in some countries including Australia, New Zealand and Argentina as well as in Western Europe where longer growing seasons prevail. Emergence of confinement systems has been paralleled by the development of specialized management strategies needed to support the higher milk yield potential of cows which has increased significantly in the US for example, from 3,191 kg in 1960 to 8,263 kg in 2000 [1].

Interest in the pasture-based dairy system in temperate regions relates to the increasing feed cost as well as concerns over environmental and animal-welfare issues associated with confinement of cows [11]. Although higher milk production is achieved by confining cows, the higher profitability and lower environmental impact of pasture-based systems result from reduced use of purchased inputs including grain, as well as less constructed infrastructure [14]. Rotationally grazed forage also provides the cheapest source of energy for dairy cows. A review of data from several countries (US, Denmark, Germany, The Netherlands, UK, France, Ireland, Australia and New Zealand) showed that on average with every 10% increase in the amount of grazed forage in the diet, the cost of milk production declined by 2.5 cents per liter [11]. These data also show that the cost of milk production in Australia and New Zealand in systems where pasture accounts for about 85-90 % of the dairy ration is about one third of the production cost of confinement systems in the US [11]. A review of pasture systems in the US showed that milk production increased linearly when cows are fed up to 10 kg concentrate per day, with an average production response of 1 kg milk per kg of concentrate [1]. However, high fossil energy prices renew interest of milk producers in reducing concentrate use through grazing. Research of pasture systems in temperate regions has focused on increasing milk production of grazing cows profitably by using rotational grazing and limiting the use of feed supplements. Type of forage and amount of energy supplement in a dairy ration have been related previously to differences in the healthfulness of milk fat [4, 7], a phenomenon which remains to be exploited by the dairy industry.

Milk Fatty Acids and Their Health Effects

Milk fat typically has a high content of saturated fatty acids (SFAs; 60-70%) followed by monounsaturated fatty acids (MUFAs; 25-30%), and has a low content of polyunsaturated fatty acids (PUFAs; ~ 5%) [9]. The milk FAs are either synthesized *de novo* in the mammary gland from acetate and butyrate or are taken-up by the mammary as pre-formed FAs from blood (originating from diet and adipose tissue).

De novo synthesized FAs include short (4:0-8:0) and medium (10:0-14:0) chain SFAs as well as about 50 % of 16:0 (palmitic acid; PA) [31]. The remaining milk FAs (saturated and unsaturated FAs with 16 to 24 carbons) originating primarily from the diet are extracted by the mammary from circulating lipoproteins [31]. Due to extensive ruminal biohydrogenation (RBH) of dietary unsaturated FAs (UFAs), the majority of FAs taken up by the mammary are saturated. However, feeding management affects mammary uptake of some dietary UFAs as well as RBH intermediates especially *trans* 18:1 and the conjugated 18:2 isomers [4]. Furthermore, SFAs from both the blood and *de novo* synthesis can be desaturated via mammary Δ-9 desaturase to their respective *cis*-9 MUFAs before esterification in milk TAGs. Mammary Δ-9 desaturase activity accounts for approximately 60% of OA present in milk fat, as well as 64-97% of RA [31].

Milk and dairy products are major dietary sources of SFAs in developed countries [22]. During the last three decades, concerns regarding association of SFAs with cardiovascular disease (CVD) have reduced consumption of milk fat in these countries [26]. The SFAs in bovine milk range from 4 to 24 carbons in length with the major ones (% of milk FAs) consisting of: PA (22-35%), 14:0 (myristic acid, 8-14%), 18:0 (stearic acid; SA, 9-14%), 12:0 (lauric acid, 2-5%), 10:0 (capric acid, 2-4%), 8:0 (caprylic acid, 1-3%), 6:0 (caproic acid, 1-5%) and 4:0 (butyric acid, 2-5%) [20]. Three SFAs in particular (palmitic, myristic and lauric) are hypercholesterolemic and increase the risk of CVD [33]; however, despite the presence of these FAs in milk fat, there is a lack of epidemiological evidence that consuming milk fat increases the risk of CVD [16, 22]. In fact, a number of epidemiological studies suggest that increased consumption of milk and dairy products may reduce the risk of CVD [22]. Although the mechanism remains unclear, the effect could be due to a reduction in blood pressure [16] and anti-atherogenic effects of the OA in milk [17].

The short chain SFAs (4:0-10:0) and SA do not raise blood cholesterol [33]. In fact, some of these FAs, such as butyric, caprylic and capric acids have health benefits. Dairy products are the major dietary source of butyric acid, which has been shown to have anti-carcinogenic properties in animal models and cancer cell lines [32]. Caprylic and capric acids provide an immediate energy source in clinical nutrition because of their rapid absorption [15].

Feeding strategies that reduce the milk fat content of the atherogenic FAs could reduce the potential adverse effects of milk fat consumption on vascular health [16, 32]. Consumption of dairy products in which 12:0, 14:0 and PA were partially replaced by healthful MUFAs and PUFAs via changes in feeding management of cows lowered total and LDL cholesterol in humans [28]. On the other hand, feeding PA rich fat supplements has become a common practice in today's dairy industry to increase energy intake and milk fat content [25]. Unfortunately, the detrimental effects of this practice on the healthfulness of milk fat by increasing milk content of PA [6] have been over looked.

Oleic acid is the dominant UFA in milk and the second major FA (after PA) constituting 20-30 % of total milk FAs [20]. Milk and dairy products are considered a major dietary source of this FA [22] which has been shown to improve vascular health. Increased dietary intake of OA at the expense of SFAs can reduce the risk of CVD by lowering blood cholesterol [33]. Similarly, consumption of nutritionally modified milk and dairy products with higher content of OA can improve vascular health [28].

The *trans* 18:1 FAs are the second major milk MUFA constituent (after OA) and can range from 0.88 to 7.65 % of total milk FAs [7] depending on diet and feeding practices. The *trans* 18:1 isomers in ruminant fat include *trans*-4 18:1 to *trans*-16 18:1 with VA being the dominant isomer (60-80% of the total *trans* 18:1) [32]. In contrast to milk, *trans*-9 and *trans*-10 18:1 are the major *trans* FAs in partially hydrogenated vegetable oils and are considered detrimental [32]. Vaccenic acid can be converted to RA in animal tissues via Δ-9 desaturase and; thereby, may have anti-carcinogenic properties [5, 24]. Furthermore, VA has been shown to have anti-atherogenic properties in mice [2]. A number of studies have shown that the consumption of butter naturally enriched with VA and RA and containing reduced levels of SFAs improved blood lipid profile and reduced CVD risk in animal models compared to standard butter [32]. However, since these effects were not as consistent in humans, more studies are needed to confirm the effects of VA and RA along with reduced levels of SFAs on the human blood lipid profile [32].

The content of PUFAs in milk fat is very low due primarily to extensive RBH. The main PUFAs in bovine milk fat are linoleic acid (LA; 18:2 n-6, 1-3% of total FAs) and ALA (0.5-2% of total FAs) [20]. Milk fat is a unique source of conjugated linoleic acid (CLA) in the human diet contributing about 75% of CLA intake [23]. The content of CLA in bovine milk ranges from 0.34 to 1.07 % of total fat depending on feeding management [10]. To date, about 20 CLA isomers have been identified in milk fat of which RA constitutes 66 -

89% [32]. During the last two decades, several health benefits have been attributed to CLA including reductions in the rates of cancer, obesity, atherogenesis, and osteoporosis [29]. However, the majority of these health claims originate from *in vitro* and animal model studies where high doses of synthetic CLA used may not simulate the typical human intake from milk and dairy products [12, 27]. The intake of total CLA from natural sources in the US (ruminant milk and meat) is estimated to be 151–212 mg/day [27]. However, an intake of 3 g/day of CLA may be needed for an equivalent anti-carcinogenic effect in human adults (70 kg body weight) [18].

As well as differences related to amount of CLA used, synthetic preparations of CLA used in animal model studies typically contain almost equal amounts of the two predominant isomers, RA and *trans*-10, *cis*-12 CLA, whereas the concentration of the later is very low in milk fat (0.01-1.61% of total CLA isomers) [32]. The biological activities reported for CLA are based on the combined effect of these two isomers [12, 27]. There is limited information on the specific effects of individual CLA isomers including RA and *trans*-10, *cis*-12 CLA which seem to have equivalent anti-carcinogenic properties [29]. The *trans*-10, *cis*-12 CLA has potent anti-lipogenic properties in rodents [29] and causes milk fat depression in dairy cows [31], whereas RA is more effective in improving the blood lipid profile in animal models and possibly humans [32]. More research on isomer specific health effects of CLA is needed.

Modification of Milk Fatty Acids through Cow Nutrition

Interest in modifying the fat content and FA composition of animal products to improve their healthfulness has been increasing over the last three decades. Nutrition and genetic selection are used to modify the fat content of milk, while the FA composition of milk has been largely ignored at the farm level. Generally the focus of modification of fat composition of milk has been to increase the content of one or more PUFA. However, RBH reduces the concentration of PUFA in milk. As a result of RBH, the profile of FAs leaving the rumen is very different from that of ingested FAs. While LA and ALA are naturally the predominant FAs in ruminant diets, SA is the main FA leaving the rumen for absorption in the small intestine. Numerous RBH intermediates (e.g. CLA and *trans* 18:1 isomers) escape complete RBH [31] and are assimilated causing a high degree of complexity of milk fat, with more than 400 FAs present [20].

The RBH of unprotected UFAs exceeds 80% [13, 19]. Over the past three decades methods have been developed to protect dietary UFAs against RBH to improve their transfer to ruminant products. The approaches used for ruminal protection of supplementary UFAs include encapsulation inside a microbial resistant shell (enrobing with formaldehyde treated protein or saturated lipids), transformation of FAs to derivatives (e.g. calcium salts or amides), or feeding whole oilseeds. However, for the majority of dietary UFAs, none of these approaches provides complete protection from RBH [19]. Thus more research is needed to improve the effectiveness of ruminal protection. Concurrently attempts have been made to increase ruminal outflow and enrich product content of RBH intermediates, particularly VA and RA by supplementing unprotected PUFAs. The potential to enrich milk with these FAs is much greater than that of dietary PUFAs [4, 23]. Supplementation of diets with LA or ALA is the most common method used to enhance the concentrations of VA and RA in ruminant products. A number of studies have shown that feeds rich in LA (e.g. sunflower, soybean) or ALA (e.g. linseed, fresh grass) markedly increase the concentrations of VA and RA in milk [4, 10].

Dairy Systems and Milk FA Composition

As well as having environmental, cost and welfare benefits, milk from pasture systems has a more healthful FA composition than that from confinement systems in terms of both the SFA: UFA ratio as well as the nature of these FAs (Table 1). Several studies have shown that compared to cows fed conserved forages in confinement, grazing cows and cows fed fresh forage produce milk with a lower content of atherogenic FAs (12:0-16:0), particularly PA, and a higher content of OA, VA, RA and ALA due to the higher content of PUFAs in fresh forage (Table 2). The predominant FA in fresh forages is ALA (50-75% of total FAs) [8]. However, PUFAs (especially ALA) are oxidized during the preservation of forage [7, 21] resulting in about 10 times less FA in hay and grass silage than that of fresh forage (Table 2). Corn silage and grain based concentrates, the main components of TMRs, have a lipid fraction that is relatively low in FA with predominant FAs being OA and LA (Table 2).

Table 1. The content of selected FAs[1] (g/100 g total FAs) and SFAs/UFAs[2] in milk from cows grazing or fed fresh forage vs. cows fed conserved forages in confinement

Fatty acid	Pasture/fresh forages[3]	Conserved forages[3]
Palmitic	25.7±1.6	31.2±3.4
Oleic	22.4±3.6	18.6±2.6
Vaccenic	3.5±0.6	1.3±0.4
α-linolenic	0.9±0.2	0.7±0.2
Rumenic	1.4±0.2	0.5±0.2
SFAs/UFAs[2]	1.8±0.3	2.7±0.5

[1]Fatty acids; [2]Ratio of saturated to unsaturated fatty acids; [3]Mean and standard deviation (SD) of 10 studies using fresh forages or pasture and 12 studies using silages and hay (calculated from the review by Kalac and Samkova, 2010 [21]).

Commercially available lipid supplements which are comprised mainly of saturated lipids rich in SA and PA, are fed largely to boost milk and fat production but also alter milk fat composition. Oilseeds are typically high in OA and LA with the exception of flaxseed, which is rich in ALA (Table 2) and are fed as an energy source, to improve reproductive performance and milk FA composition, and to suppress enteric methanogenesis.

The differences in milk FA composition between pasture and confinement production systems have been attributed mainly to the higher intake of ALA in pasture systems and its ruminal metabolism which results in increased ruminal outflow of ALA and RBH products, particularly VA and SA (Figure 1).

Since conversion of VA to SA is the last step of RBH and is rate limiting, these FAs comprise the main substrates for mammary Δ-9 desaturase. Therefore, increased uptake of VA and SA by the mammary gland elevates the milk concentrations of RA and OA respectively, as well as VA and SA (Figure 1). Furthermore, the increased mammary uptake of circulating PUFAs (e.g. ALA) and their RBH intermediates (e.g. *trans* 18:1 FAs) reduces the milk content of *de novo* synthesized FAs, in particular PA, through the inhibition of mammary enzymes involved in *de novo* FA synthesis [4].

Besides the high intake of ALA in pasture systems, the unique ruminal digestion condition caused by the highly fermentable structural carbohydrates in high quality pasture relates to a lower ruminal retention time [30] and plays a role in causing differences in milk FA composition from that of confined cows.

Table 2. Ether extract (EE), total fatty acid (FA) concentration and FA composition of fresh and conserved forages, grains, oilseeds and commercial lipid supplements used in dairy rations

	EE	FA	FA composition (g/100 g total FAs)				
	(%DM[1])	(%DM)	Palmitic	Stearic	Oleic	Linoleic	ALA[2]
Fresh forage							
Perennial ryegrass[3]	3.6	1.8	12.9	3.0	3.2	14.6	65
Meadow fescue[3]	3.2	1.6	17.7	1.5	4.4	15.9	43.4
Tall fescue[4]	-	2.2-3.9	11-22	1.4-4.9	2-4.	9.1-11.8	55-72
Cocksfoot[3]	3.6	1.9	11.2	2.6	-	76.5	-
Timothy[4]	-	2.0-2.1	19.5	4.3-5.1	5-6	15	50
Red clover[3]	5.1	2.3	14.2	3.7	-	5.6	72
White clover[3]	5.0	2.2	6.5	0.5	6.6	18.5	61
Alfalfa[3]	5.3	2.5	2.7	<0.5	<0.5	1.7	95
Perennial sward[5]	3.6-4.3	-	11.0	1.4	2.0	14.5	46.5
Hay							
Cocksfoot[4]	-	0.2	24	2.8-4.5	3-4	16-17	27-35
Perennial ryegrass[4]	-	0.3	15.8	1.8	2.0	14.0	55.9
Alfalfa[4]	-	0.1	30.0	6.0	8.0	24.4	23.2
Perennial sward[5]	2.4	-	16.9	2.5	4.8	13.3	26.1
Silage							
Corn[4]	-	0.1-0.4	16-17	2.4-2.9	19-24	49	3-11
Perennial ryegrass [4]	-	0.4	21.2	2.0	2.8	13.4	52
Perennial sward[5]	3.7	-	13.1	1.3	4.1	21.2	29.8
Grains							
Corn[3]	3.6-5.0	3.2	16.3	2.6	30.9	47.8	2.3
Barley[3]	1.7-2.9	1.6	27.6	1.5	20.5	43.3	4.3
Oilseeds							
Sunflower[3]	38-45	34.7	5.5	3.6	21.7	68.5	<0.5
Soybeans[3]	16-25	19	10.7	3.9	22.8	50.8	6.8
Flaxseeds[6]	36	-	5.1-7.1	2.8-4.0	16-22	15-18	50-59
Commercial lipid supplements [7]							
Alifet-High energy®	99	-	22.3	31.7	11.5	0.19	0.16
Energy Booster®	100	-	41.9	44.5	5.5	1.7	0.72
Megalac®	85	-	46.0	4.48	35.3	8.2	0.30

[1]Dry matter; [2]α-linolenic acid; [3]Schroeder et al., 2004 [30]; [4]Kalac and Samkova, 2010 [21]; [5]Vahmani et al., unpublished data; [6]Woods et al., 2009 [34]; [7]Carriquiry et al., 2008 [3].

Grazing cows are also typically fed an energy supplement consisting of non- structural carbohydrates (NSC) in discrete meals which in consort with low intake of effective fiber results in the potential for an episodic reduction in ruminal pH in grazing cows compared to confined cows consuming TMR.

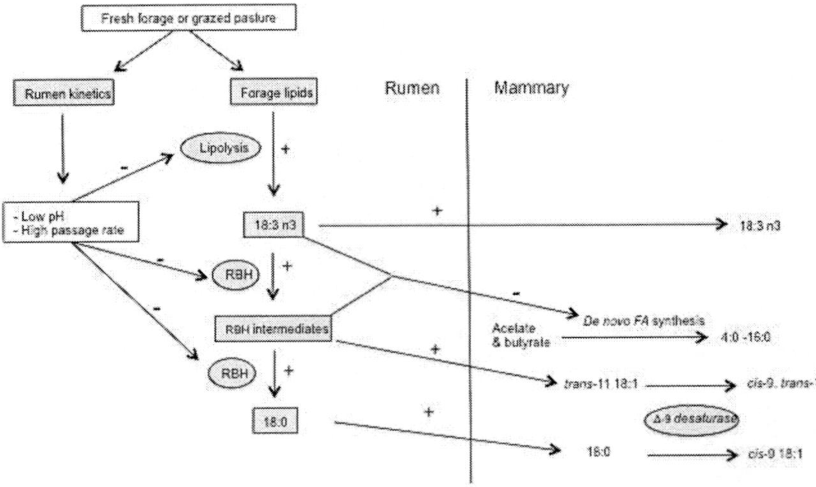

Figure 1. The relationship between ruminal lipid metabolism and milk fatty acid composition in grazing cows. 18:3 n3 = α-linolenic acid, RBH = ruminal biohydrogenation, 4:0 = butyric acid, 16:0 = palmitic acid, *trans*-11 18:1 = vaccenic acid, *cis*-9, *trans*-11 18:2 = rumenic acid, *cis*-9 18:1 = oleic acid.

These conditions (low ruminal pH and ruminal retention time) can limit both lipolysis (the prerequisite for RBH) as well as RBH of dietary FAs (Figure 1). As a result, more dietary UFAs (e.g. intact ALA) and RBH intermediates (e.g. VA) exit the rumen before their complete hydrogenation to SFAs (Figure 1). Furthermore, some metabolites in fresh forages (e.g. saponins, polyphenols and catecholamines) can have inhibitory effects on ruminal lipolysis and biohydrogenation [21]. Lipoxygenases are activated when the cells of growing plants are disrupted by chewing releasing some volatile FA oxidation products (e.g. hydroperoxides, alcohols, aldehydes and ketones), which have antimicrobial properties and may reduce lipolysis and RBH. These processes increase the ruminal outflow of ingested UFAs and their RBH intermediates in grazing cows [7].

We recently performed a direct, controlled comparison of pasture and confinement systems on milk FA composition. Our unpublished data (Figure 2) showed that milk from early to mid lactation grazing Holstein cows had a significantly ($P<0.05$) lower SFA to UFA ratio (1.9 vs. 2.9) compared to that of confined cows, reduced concentration (% of milk total FAs) of PA (29.3 vs. 34.9) and higher concentrations of OA (20.8 vs. 15.5), VA (2.9 vs. 0.98), RA (0.91 vs. 0.32) and ALA (0.63 vs. 0.43).

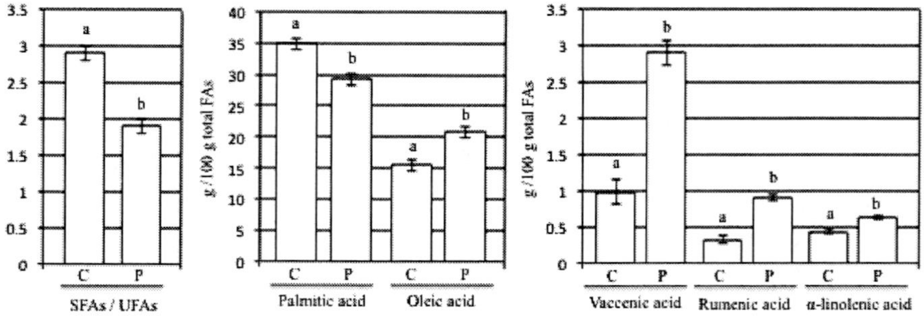

Figure 2. Effect of dairy system on the ratio of saturated to unsaturated fatty acids (SFAs/UFAs) and concentrations of selected fatty acids (FAs) in milk. Values within each fatty acid with different letters are significantly different ($p < 0.05$). C = confinement system including data from 8 cows fed a total mixed ration containing (DM basis) 56% forage (28:22:6 mix of grass silage, corn silage, grass hay) and 44% concentrate; P = pasture system including data from 8 cows grazed on a mixed sward pasture under rotational management and supplemented with concentrate at the rate of 0.25 milk yield to a maximum of 8 kg as fed/day.

These results are in agreement with the indirect comparisons from previous studies (Table 1) confirming that the milk from pasture systems has a significantly healthier FA profile compared to the milk from confinement systems.

CONCLUSION

Pasture systems in temperate regions produce milk with a healthier FA profile than that of confinement systems. Direct and indirect comparisons show that compared to confinement systems, milk produced in pasture systems has a higher content of beneficial UFAs (OA, VA, RA and ALA) and lower content of harmful SFAs, particularly PA.

The use of pasture systems in temperate North America could improve the image of milk fat which is typically known to have a high content of SFAs and low content of PUFAs. Unfortunately dairy farmers are currently not rewarded for producing nutritionally improved milk fat.

References

[1] Bargo, F., Muller, L. D., Kolver, E. S., Delahoy, J. E. Invited review: production and digestion of supplemented dairy cows on pasture. *J. Dairy Sci.* 2003 Jan.;86(1):1-42.

[2] Bassett, C. M., Edel, A. L., Patenaude, A. F., McCullough, R. S., Blackwood, D. P., Chouinard, P. Y., et al. Dietary vaccenic acid has antiatherogenic effects in LDLr-/- mice. *J. Nutr.* 2010 Jan.;140(1):18-24.

[3] Carriquiry, M., Weber, W. J., Baumgard, L. H., Crooker, B. A. In vitro biohydrogenation of four dietary fats. *Anim. Feed Sci. Technol.* 2008 Apr. 141(3-4):339-355.

[4] Chilliard, Y., Glasser, F., Ferlay, A., Bernard, L., Rouel, J., Doreau, M. Diet, rumen biohydrogenation and nutritional quality of cow and goat milk fat. *Eur. J. Lipid Sci. Technol.* 2007 Aug.;109(8):828-855.

[5] Corl, B. A., Barbano, D. M., Bauman, D. E., Ip C. *cis*-9, *trans*-11 CLA derived endogenously from *trans*-11 18:1 reduces cancer risk in rats. *J. Nutr.* 2003 Sep.;133(9):2893-2900.

[6] DeLand, K. E., Lock, A. L., Preseault, C. L., Allen, M. S. Effect of feeding a C16:0-enriched fat supplement on milk fatty acid composition. In: ADSA-ASAS Joint Annual Meeting ; 2011 Jul 10-14; New Orleans, Louisiana, US. *J. Dairy Sci.* 94 (E-Suppl.). 2011. p.212.

[7] Dewhurst, R. J., Shingfield, K. J., Lee, M. R. F., Scollan, N. D. Increasing the concentrations of beneficial polyunsaturated fatty acids in milk produced by dairy cows in high-forage systems. *Anim. Feed. Sci. Technol.* 2006 Dec.;131(3-4):168-206.

[8] Dewhurst, R. J. Forage Lipids and Effects on Ruminant Productivity. In: Timmis, K. N., editor. *Handbook of Hydrocarbon and Lipid Microbiology*. Springer-Verlag Berlin Heidelberg; 2010. p. 3160-3165.

[9] Dewhurst, R. J. Targets for milk fat research: Nutrient, nuisance or nutraceutical? *J. Agric. Sci.* 2005;143(5):359-367.

[10] Dhiman, T. R., Nam, S. H., Ure, A. L. Factors affecting conjugated linoleic acid content in milk and meat. *Crit. Rev. Food. Sci. Nutr.* 2005; 45(6):463-482.

[11] Dillon, P. Achieving high dry-matter intake from pasture with grazing dairy cows. In: Elgersma, A., Dijkstra, J., Tamminga, S. (eds.). Fresh herbage for dairy cattle: the key to a sustainable food chain. Springer, Dordrecht, The Netherlands; 2006. p. 1-26.

[12] Dilzer, A., Park, Y. Implication of Conjugated Linoleic Acid (CLA) in Human Health. *Crit. Rev. Food Sci. Nutr.* 2012;52(6):488-513.

[13] Doreau, M., Chilliard, Y. Effects of ruminal or postruminal fish oil supplementation on intake and digestion in dairy cows. *Reprod. Nutr. Dev.* 1997;37(1):113-124.
[14] Fontaneli, R. S., Sollenberger, L. E., Littell, R. C., Staples, C. R. Performance of lactating dairy cows managed on pasture-based or in free stall barn-feeding systems. *J. Dairy Sci.* 2005 Mar.;88(3):1264-1276.
[15] Gibson, R. A. Milk Fat and Health Consequences. Nestle Nutr. Workshop Ser Pediatr Program. Nestlé Nutr. Inst. Workshop Ser. Pediatr. Program. 2011;67:197-207. doi: 10.1159/000325585. Epub. 2011 Feb. 16.
[16] Givens, D. I. Food chain and health. Milk in the diet: good or bad for vascular disease? *Proc. Nutr. Soc.* 2012 Feb. 2012;71(1):98-104.
[17] Haug, A., Hostmark, A. T., Harstad, O. M. Bovine milk in human nutrition-a review. *Lipids Health Dis.* 2007 Sep;6:25.
[18] Ip, C., Singh, M., Thompson, H. J., Scimeca, J. A. Conjugated linoleic acid suppresses mammary carcinogenesis and proliferative activity of the mammary gland in the rat. *Cancer Res.* 1994 Mar.;54(5):1212-1215.
[19] Jenkins, T. C., Bridges, W. C. Protection of fatty acids against ruminal biohydrogenation in cattle. *Eur. J. Lipid Sci. Technol.* 2007;109(8):778-789.
[20] Jensen, R. The composition of bovine milk lipids: January 1995 to December 2000. *J. Dairy Sci.* 2002 Feb.;85(2):295-350.
[21] Kalač, P., Samková, E. The effects of feeding various forages on fatty acid composition of bovine milk fat: A review. *Czech. J. Anim. Sci.* 2010; 55(12):521-537.
[22] Kliem, K. E., Givens, D. I. Dairy Products in the Food Chain: Their Impact on Health. *Annu. Rev. Food Sci. Technol.* 2011;2:21-36.
[23] Lock, A. L., Bauman, D. E. Modifying milk fat composition of dairy cows to enhance fatty acids beneficial to human health. *Lipids* 2004 Dec.;39(12):1197-1206.
[24] Lock, A. L., Corl, B. A., Barbano, D. M., Bauman, D. E., Ip, C. The anticarcinogenic effect of *trans*-11 18:1 is dependent on its conversion to *cis*-9, *trans*-11 CLA by delta9-desaturase in rats. *J. Nutr.* 2004 Oct.;134 (10):2698-2704.
[25] Lock, A. L., Preseault, C. L., DeLand, K. E., Allen, M. S. Feeding a C16:0-enriched fat supplement increased the yield of milk fat and improved feed efficiency. In: ADSA-ASAS Joint Annual Meeting; 2011 Jul. 10-14; New Orleans, Louisiana, US. *J. Dairy Sci.* 94 (E-Suppl.). 2011. p.737.

[26] Lubary, M., Hofland, G. W., ter Horst, J. H. The potential of milk fat for the synthesis of valuable derivatives. *Eur. Food Res. Technol.* 2011; 232(1):1-8.

[27] McCrorie, T. A., Keaveney, E. M., Wallace, J. M. W., Binns, N., Livingstone, M. B. E. Human health effects of conjugated linoleic acid from milk and supplements. *Nutr. Res. Rev*. 2011 Dec.;24(2): 206-227.

[28] Noakes, M., Nestel, P. J., Clifton, P. M. Modifying the fatty acid profile of dairy products through feedlot technology lowers plasma cholesterol of humans consuming the products. *Am. J. Clin. Nutr*. 1996 Jan.;63 (1): 42-46.

[29] Park, Y. Conjugated linoleic acid (CLA): Good or bad *trans* fat? *J. Food Comp. Anal.* 2009 Dec.; 22S: S4–S12.

[30] Schroeder, G., Gagliostro, G., Bargo, F., Delahoy, J., Muller, L. Effects of fat supplementation on milk production and composition by dairy cows on pasture: a review. *Livest. Prod. Sci*. 2004 Mar.;86:1-18.

[31] Shingfield, K. J., Bernard, L., Leroux, C., Chilliard, Y. Role of trans fatty acids in the nutritional regulation of mammary lipogenesis in ruminants. *Animal* 2010 Jul.;4(7):1140-1166.

[32] Shingfield, K. J., Chilliard, Y., Toivonen, V., Kairenius, P., Givens, D. I. *Trans* fatty acids and bioactive lipids in ruminant milk. *Adv. Exp. Med. Biol.* 2008;606:3-65.

[33] Williams, C. Dietary fatty acids and human health. *Ann. Zootech*. 2000 May-Jun.;49(3):165-180.

[34] Woods, V. B., Fearon, A. M. Dietary sources of unsaturated fatty acids for animals and their transfer into meat, milk and eggs: A review. *Livest. Sci.* 2009;126(1-3):1-20.

In: Milk Fat
Editor: Linda. B. Rudd

ISBN: 1-978-62618-353-7

Chapter 4

Milk Fat Globules' Microstructure

Nicoletta Fucà*, Gaetano Impoco†
and Giuseppe Licitra
CoRFiLaC, Italy
DISPA, University of Catania, Italy

Abstract

Milk fat globules are subject to several changes due to milk industrial processes and storage conditions. There is a potential industrial interest in monitoring changes in milk microstructure subject to various sanitation treatments and different storage temperatures. Most variations are related to the milk fat globules membrane (MFGM), whose main role is to keep isolated fat globules and protect them from lipolysis. Homogenization and heat treatments favour the deposit of milk proteins on the MFGM and their interactions, thus facilitating aggregation and fusion, and variation of size, shape and distribution of globules. Fat globules' structural properties can be studied by several techniques, such as X-ray diffraction and differential scanning calorimetry, scanning electron, fluorescence, and confocal microscopy, light scattering, SDS PAGE, and image analysis of milk micrographs. We review a number of techniques used to study the effects of industrial treatments on fat microstructure, aiming to detect measurable differences between strong and weak treatments.

Keywords: milk fat, MFGM, microstructure, sensing technologies

*E-mail address: fuca@corfilac.it
†E-mail address: impoco@corfilac.it

1. Introduction

Milk fat globules are made of a core of triglycerides surrounded by a membrane (*milk fat globule mambrane*, MFGM) composed by proteins, phospholipids, trace elements, enzymes, and carotene [19]. Some of the components are emulsifiers and prevent individual globules from joining together. MFGM protects fat globules from lipolysis catalysed by lipase activity, mainly linked to the casein micelles.

Milk fat microstructure goes through variations caused by industrial treatments (i.e., mechanical and thermal treatments). Homogenisation is the mechanical process that results in an outstanding reduction of fat globule size due to shear stress, inertial forces and cavitation. According to Stoke's law, creaming rate decreases with globule size due to the density difference between milk fat and the aqueous phase [21]. Homogenisation does not ensure milk safety, thus heat treatments are performed to preserve microbiological quality, extending milk shelf-life [12].

Several authors have studied the effects of industrial treatments using different technologies, ranging from spectroscopy to microscopy. This chapter tries to give an overall view on milk fat globule features and on microstructural changes occurring with thermal or mechanical stress. We do not venture on trying to give a comprehensive review of the huge amount of work that has been done on this topic. Some important findings are discussed and their role is highlighted to describe some aspects of the problem.

In Section 2. we focus on a few techniques that have been successfully applied to the study of milk fat. A brief introduction on each presented technique is associated with the description of its uses for this problem. Section 3. delves into milk fat properties and their changes due to thermal and mechanical stresses, e.g., due to industrial treatments.

2. Techniques

Table 1 summarises the properties we examine in this chapter and the techniques used to study those features. A number of techniques have been employed to understand microstructural properties of the milk fat phase. In the following, some of these methods are introduced, together with some applications to the study of milk fat.

Table 1. Milk fat structural properties affected by industrial treatments and the main technologies used to detect their changes: X-ray diffraction (XRD), differential scanning calorimetry (DSC), microscopy/image analysis, light scattering, gel electrophoresis, chromatography, mass spectrometry, photon density wave spectroscopy (PDW).

	MFGM thickness	melting crystallization	structure & size	MFGM components	distribution aggregation, fusion
XRD	Michalski et al. [22]				
DSC		Michalski et al. [22]			
microscopy & image analysis			Gallier et al [9] Ong et al. [24] Fucà et al. [8] Keenan et al. [16] Bermúdez-Aguirre et al. [1] Impoco et al. [14]	Evers et al. [6]	Fucà et al. [8]
light scattering			Cano-Ruiz & Richter [3] Lee & Sherbon [18] Michalski et al. [20] Huppertz et al. [13] Michalski et al. [22] Ye et al. [35] Ong et al. [24]		
gel electrophoresis				Keenan et al. [16] Lee & Sherbon [18] Ye et al. [37] Ye et al. [35]	
chromatography				Keenan et al. [16]	
mass spectroscopy				Gallier et al [9]	
PWDS		Vargas Ruiz et al. [29]			

2.1. X-ray diffraction

X-ray diffraction (XRD) is a non-destructive technique to analyse structural, chemical and physical properties of materials. The scattered intensity of an X-ray beam hitting a sample is observed by measuring incident and scattered angle, polarization, and wavelength or energy. From these measurements, chemical composition, size and structural features can be computed.

Michalski et al. [22] employed XRD to measure thickness of the bilayer and trilayer structures of fat globules in milk. Cream and anhydrous milk fat samples were stabilised at 60°C and then cooled down to observe thickness variations of small and large fat globules with respect to temperature. Temperature was then raised again to 60°C and observations repeated. Measurements were taken at regular intervals during the cooling process.

2.2. Differential scanning calorimetry

Differential Scanning Calorimetry (DSC) is used to investigate the stability of biological systems. DSC directly measures heat changes that occur in biomolecules during controlled increase or decrease in temperature. The sample and a reference are subject to controlled temperature changes and heat flux of both sample and reference is compared. Thus, structural changes occurring at interest points, such as melting and crystallisation, can be studied.

Michalski et al. [22] employed DSC in conjunction with XRD (see Section 2.1.) to observe crystallization properties of milk fat globules. They focussed on thickness variations (using XRD) at melting and crystallisation points, detected using DSC.

2.3. Microscopy

Interest in the microstructure of dairy products has been growing steadily in the last few decades. At the same time, different microscopy techniques have been developed to visualize nano-features of food products. Microscopy has been employed as a complementary method for quality assessment, defect detection and product development. Acquired micrographs provide insights on size, shape and arrangement of food components.

2.3.1. Optical techniques

Fluorescence microscopy is an optical technique based on the fluorescence phenomenon. A fluorescent dye is added to the specimen and excited using a light source. Emitted light is captured by an optical system. Several dyes are available each binding to a specific target molecule class. Fluorescence microscopy is employed in food science to highlight different structural elements.

The acquisition system of fluorescence microscopes is basically a standard optical system. It thus exhibits all focusing limitations of optical microscopes, such as severe blurring. Confocal Laser Scanning Microscope (CLSM) is based on the same physical principles of fluorescence microscopy. However, it differs from conventional fluorescence microscopy in the optical system. Confocal optics is based on the *pinhole* principle: lens aperture is extremely small. The laser beam is thus highly focussed. Due to the rejection of out-of-focus light contributions, higher resolution can be obtained. Small aperture allows only a small area to be imaged. The full image is obtained by scanning the specimen surface in a rectangular pattern.Three-dimensional stacks can be imaged by acquiring multiple slices at different depths.

In milk samples, fluorescence has been employed to discriminate fat globules' core from the MFGM. Evers et al. [6] developed a staining technique to mark proteins and phospholipids, providing information about MFGM components. This information was used to observe structural and compositional differences in the MFGM among human, ovine and bovine milk. Both standard fluorescence microscope and CLSM were used. Gallier et al. [9] studied structural changes of the MFGM caused by industrial treatments, using CLSM. Three dyes where used to stain proteins, triglycerides and phospholipids in raw and processed milk cream. Staining allowed to detect phospholipid loss, globule size variation, and adsorption of milk proteins to the MFGM. Effects of treatments were also studied in [24, 8]. Ong et al. [24] evaluate treatments in the context of the cheese making process. Morphological and size changes were observed both for globule fat core and for MFGM. Fucà et al. [8] focussed on fat core, analysing several microstructural properties, such as size, aggregation, fusion and distribution uniformity of droplets.

2.3.2. Electron techniques

Milk treatments have also been studied using electron microscopy. This technology provides magnification factors of several orders of magnitude higher than optical techniques. An electron beam is focussed onto the sample surface in the same way as in optical scanning techniques. Electrons are deflected using a sort of magnetic lens, in a high-vacuum chamber. Scanning Electron Microscopy (SEM) and Transmission Electron Microscopy (TEM) only differ in the way electrons are detected after hitting the specimen surface. SEM measures number of backscattered or secondary electrons, whereas TEM counts electrons transmitted through a thin sample. Observation of food specimens requires a specific preparation technique, since sample surface must be electrically conductive. This preparation generally affects food microstructure and might cause structural changes. Keenan et al. [16] studied structural changes of MFGM in milk subject to homogenisation using SEM, whereas Bermúdez-Aguirre et al. [1] evaluated the effects of thermosonication and heat treatments. TEM has been employed in [9] to observe protein adsorption on the MFGM after heating.

2.3.3. Image analysis

Image analysis is becoming increasingly popular due to its ability to deliver quantitative, objective and repeatable estimates. Microscopists' evaluation is never fully objective and conclusions could be biased. Hence, image analysis is often used in conjunction with microscopy techniques. Unfortunately, the quality of computer analysis of digital images is far from being comparable with human visual capabilities. Computer measurements are objective and deterministic but often affected by misinterpretation of image features. Impoco et al. [14] quantified structural properties of fat core in milk confocal micrographs, using a set of mathematical descriptors. A comparison of human visual evaluations, given by trained panellists, with fully-automatic computer measurements showed that this method reliably describes perceived structural features. This image analysis tool as been used in [8] to quantitatively describe the microstructure of the lipidic phase in milk.

2.4. Light scattering

Light scattering measures the intensity of the light scattered by the molecules in the sample, in dependence of the scattering angle. In Static Light Scattering (SLS), a laser beam illuminates the sample to be analysed. Photon detectors are used to measure the scattering intensity with respect to the scattering angle. The resulting scattering curve contains information about particle size, shape and molar mass. The sample can be measured in its native state, as long as the particle concentration is small enough to avoid multiple scattering effects. Dynamic light scattering (DLS) allows to measure particle size at the nanometric scale. Here, fluctuations of the scattered light are detected, rather than mere light intensity. Detected light acts as an imprinting of particle motion. Analysis of the fluctuation of the scattered light thus yields information about the particles. Intensity fluctuations are characterized by computing the intensity correlation function, whose analysis provides the diffusion coefficient of the particles. False measurements due to multiple scattering result in significant errors in both SLS and DLS measurements, often without ever being noticed. Light scattering techniques were used in [3, 18, 20, 13, 22, 35, 24] to estimate size and mean diameter of fat globules in milk.

Care should be taken when measuring size of milk fat globules. The temperature of milk sample affects light scattering since it induces melting and crystallization (see Section 2.8.).

2.5. Gel electrophoresis

Gel electrophoresis is a technique used to separate proteins or nucleic acids according to their charge or the length of the polypeptide chain. A denaturing anionic agent is used in Sodium Dodecyl Sulfate PolyAcrylamide Gel Electrophoresis (SDS-PAGE) to unfold proteins and impart a negative charge to the unfolded proteins. SDS-PAGE is employed to detect MFGM proteins in [16, 18, 37, 35]. Changes in protein composition on the MFGM are observed to study the effects of homogenisation and thermal treatments [18], thermal and high-pressure processing [35] or seasonal variations [37].

2.6. Chromatography

Chromatography techniques are used to separate mixture molecules. Chromatographic separation is based on the aptitude of each molecule for distributing between two different phases (stationary and mobile). Interactions between mixture and the two phases are usually secondary chemical bounds. Due to the retention properties of the stationary phase, flow speed is different for each constituent of the mixture and is related to its molecular mass and polarity. Eluted constituents are reported in a chromatogram.

In Thin Layer Chromatography (TLC) the sample solution is placed onto a thin layer of adsorbent on a flat, inert substrate. A shallow layer of solvent is added. Separation is obtained by capillarity as the solvent rises through the sample mixture. Chromatography is accurate enough to allow the separation of compounds in their constituents. Thin-layer chromatography was used in [16] to assess the phospholipidic profile in homogenised milk samples.

2.7. Mass spectrometry

Mass spectrometry measures the atomic composition in a compound. The sample is vaporized into a vacuum chamber, where particles are ionised and accelerated. A magnetic force is applied orthogonally to the direction of motion of accelerated particles. Being electrically charged, particles are deflected from linear motion and hit a detector. The deflection angle depends both on mass and charge. If charge is kept constant for all particles, deflection varies only with mass. By varying the applied magnetic force, the amount of particles for each element in the compound can be measured. Mass spectrometry was employed in [9] to profile the phospholipidic compositions of MFGM in raw and processed milk, raw cream and buttermilk powder.

2.8. Photon Density Wave spectroscopy

Photon density wave (PDW) spectroscopy characterises adsorption and scattering properties of materials, without any sampling or sample preparation. An intensity-modulated light source is coupled into the material surface, thus generating a diffuse photon density wave in the medium that propagates away from the source. In heterogeneous media, the diffusive wave is distorted in predictable ways. Distortions are measured by light detectors on the tissue surface, and the information is converted into spectroscopic maps of the material interior. PDW spectroscopy was employed in [29] to measure melting and crystallisation properties of milk fat globules. Structural changes due to temperature strongly affect light scattering properties of milk. That means that measurements rely on light scattering (e.g. particle sizing) should account for the thermal history of milk samples.

3. Fat globules properties

The microstructure of milk fat phase is affected by industrial treatments and storage temperature. Pressure and other mechanical stresses induce changes in fat globule size and shape, and in their spatial distribution, as well as in MFGM structure and composition. These changes can also be enhanced by storage conditions, e.g., storing at low temperature favours fat-crystallisation.

3.1. Effects of homogenisation

Effects of industrial treatments on the microstructure of the milk fat phase have been treated since the '60s. Jackson et al. [15] and Fox et al. [7] had already studied lipo-protein complexes developed on MFGM of homogenized milk. Later on, Mulder and Walstra [23] and Walstra [31] delved into the structural changes of milk fat globules, discussing size reduction and adsorption of active substances of milk serum (i.e., proteins) with milk fat globules surface. Fat globules and adsorbed protein constitute a bounded system with lower total energy and, as such, a more stable one. The adsorption process induce protein conformational changes [33].

A deeper insight into the distribution of MFGM components and their interactions with milk proteins is given by Keenan et al. [16]. Homogenisation causes a displacement of a portion of phospholipids and cholesterol, originally associated with fat globules. Additionally, some enzymes associated to MFGM (phosphodiesterase I, 5′-nucleotidase and acid phosphatase) are released into the serum. However, those enzymes are loosed in different percentages. Such results could indicate their affinity with the globule membrane. Electrophoresis and observations by electron microscope demonstrated that homogenized globule surface is bounded only by caseins [16, 26, 18]. Figure 1 summarizes the effects just described.

Homogenisation increases fat surface area more than 10-fold [26]. Anyway, size reduction of milk fat globules is dependent on the intensity of the homogenization process. Lower pressures give a more variable size distribution than that of strongly homogenized milk [20, 30]. Figure 2 shows typical CLSM micrographs of mild (Figure 2(a)) and strong (Figure 2(b)) homogenisation, shot by Fucà et al. [8]. Large and small globules appear in mildly homogenised samples, whereas only small droplets appear in strongly homogenized samples. Such difference is evident to human observers as well as to a computer image analysis method, as proven by Fucà et al. [8]. Here, samples subject to a mild homogenisation were grouped with raw milk, due to the large variation of globule size. On the other hand, strongly homogenised milk samples were all grouped together, being characterised by a narrow size distribution.

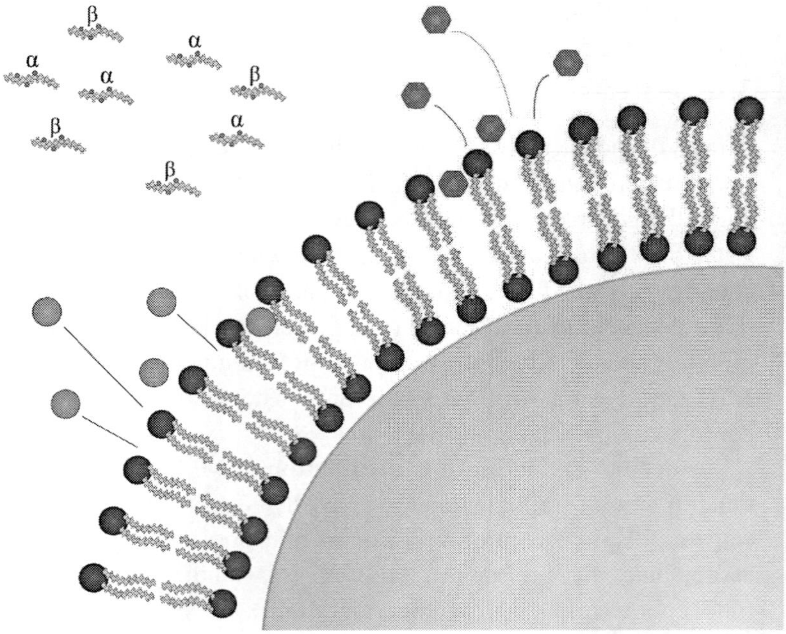

Figure 1. Schematic representation of MFGM changes upon homogenization. Orange circles represent caseins, whereas pink hexagons represent enzymes.

Recently, new industrial treatments have been proposed as alternative processes to reduce fat globule size and thus creaming of raw milk. A number of studies [27, 13, 35] inspected the effects of high-pressure homogenization. Conflicting results have been noticed by Thiebaud et al. [27] and by Huppertz et al. [13], regarding fat globules' size reduction upon high-pressure homogenisation (HP). Thiebaud et al. [27] performed two consecutive steps for a very short time. After the first step, globule size was significantly reduced even though a residual variability was maintained. After the second step, size was further decreased and its variability was significantly reduced. Hupperts et al. [13] performed only one step for a longer time. In this case, no significant difference was found. The same result has been obtained in [35], where the HP treatment performed in one step for a long time did not change the average fat globule size. Ye et al. [35] also found that HP treatment increased the amount of whey proteins bounded with the MFGM.

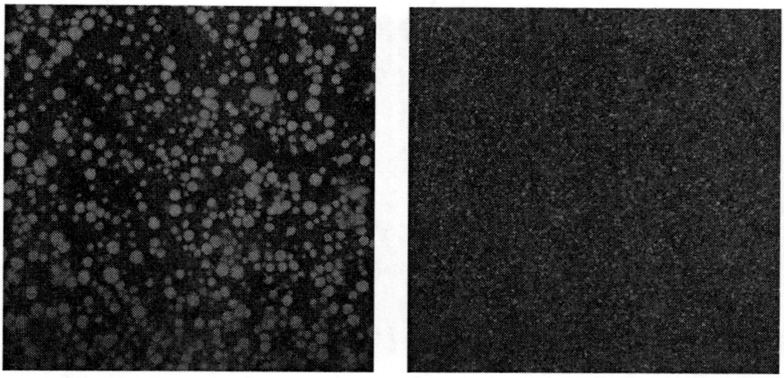

(a) Mildly homogenised. (b) Strongly homogenised.

Figure 2. Confocal micrographs of milk samples subject to different treatment. Mildly homogenisation partially reduces fat globules, whereas strong homogenisation induces smaller and equally-sized globules.

3.2. Effects of heating

Changes of fat globule structure and of MFGM can also be a consequence of heat treatments commonly used in the industry. Homogenization efficiency increases with temperature from 42°C to 72°C and stabilizes around 72°C - 77°C [12]. One of the effects of heat treatments on milk fat globules is phospholipid loss from MFGM [20]. In [9], phospholipidic profiles were measured for raw and commercial milk and cream, using tandem mass spectrometry. This technique outputs a spectrum of the distribution of molecules in a compound. Significant loss of phospholipid fractions was observed in commercial milk. This loss was balanced by milk proteins. Abundance of α-casein (αCN), β-casein (βCN), κ-casein (κCN), α-lactalbumin (αLA), and β-lactoglobulin (βLG) in *recombined fat globule membrane* (RFGM) was also studied in [25] and [3] using SDS-PAGE. Heating favours binding of serum proteins (αLA and βLG) [18]. Figure summarizes MFGM changes caused by heating.

Adsorption of proteins to the MFGM has been observed by TEM and CLSM [25, 9]. Fat globules are mainly surrounded by intact caseins, and micelle fragments (Figure 4). A thin layer around the globules is made by whey

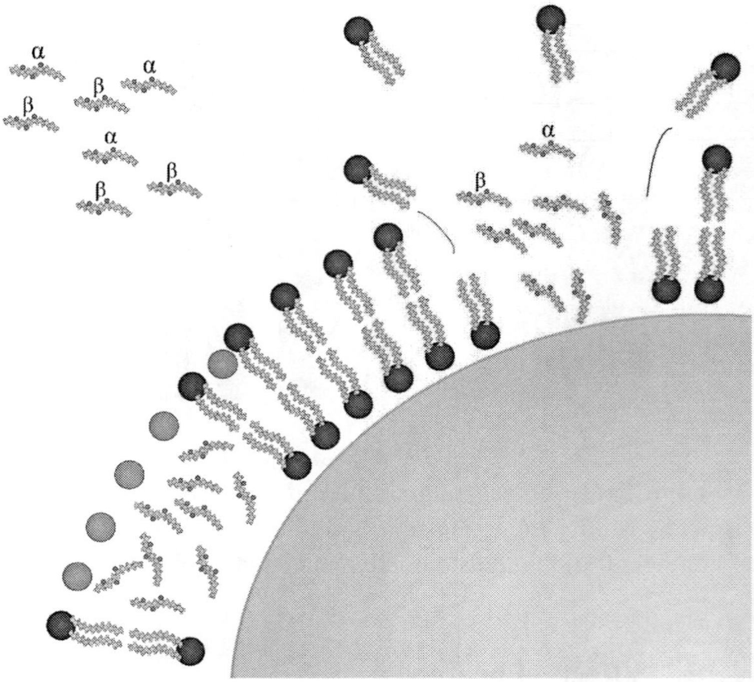

Figure 3. Schematic representation of MFGM changes upon heating. Orange circles represent caseins, whereas α and β indicate whey proteins.

proteins [25].

The extent of lipid-protein interactions is also influenced by heating temperatures and time [28, 18]. The serum proteins adsorb onto the MFGM only when fat globules are heated above 70°C [26] and their bindings gradually increase with the duration of heating [18].

Fucà et al. [8] identified two main groups of milk samples on the basis of the heating treatment. Milk samples treated at a temperature $\leq 65°$ C (*weakly treated*, WT) were discriminated from those treated at a temperature $\geq 75°$ C (*strongly treated*, ST). The differences observed between the two groups were related to globule size (variation due to the interaction of homogenisation and process temperature), and fat globules aggregation and fusion. Lower temperatures favoured more marked aggregation and fusion. The same groupings were found both by human observers (trained panellists) and using automatic mea-

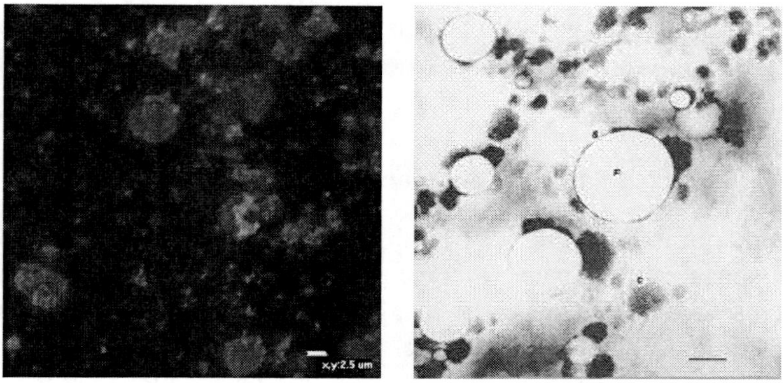

(a) CLSM. (b) TEM.

Figure 4. Micrographs of heated milk samples. Fat globules are surrounded by protein deposits, mainly made of caseins. TEM image courtesy of Sharma et al. (1996).

surements with an ad-hoc computer tool [14]. Fat globules aggregation is attributable to those changes occurring on the MFGM proteins. As already mentioned above, heating causes several changes in the MFGM proteins, such as their denaturation and interactions with whey proteins (α-LA and β-LG) via sulfhydryl-disulphide bonds [3, 5, 11, 17, 26, 36]. Both α-LA and β-LG associate with MFGM upon heat treatment and their extent increases with raising temperature up to $85°$C, remaining constant at higher temperatures [4, 36]. Caseins associated with the MFGM via disulphide (S-S) bonds between κ-casein and MFGM proteins or β-LG already associated with MFGM [35]. It is supposed that fat globules aggregation of WT milk types could be due to MFGM exposed S-S and sulfhydryl (SH) groups that can bond proteins placed on other MFGMs. In the ST (heat treatment $\geq$ 75°C) milk group, MFGM disulphide and sulfhydryl groups have already bound to β-LG/κ-CN complexes, avoiding fat globule aggregation.

Besides homogenisation, thermal treatments can be combined to other processes. This is the case of thermosonication. Milk samples is placed in a heated chamber and subject to ultrasound treatment. Supporters claim that, while ensuring product safety and homogeneity, this method retains milk sensory properties. Bermúdez-Aguirre et al. [1] studied the effects of thermosonication on milk fat microstructure. The MFGM in heated samples was less damaged compared to the sonicated ones. The membrane was disrupted in sonicated samples and triacylglycerols were released into the serum. Fat globules were smaller than $1\mu m$ and had a coarse surface due to casein adsorption.

3.3. Effects of storage temperature

Stability of milk products is partially dependent on the storage conditions. The maintenance of their macroscopic properties reflect what is happening at the microscopic level. One of the changes due to the storage temperature is phase transition of milk fat globules (i.e. melting and crystallisation). The globules are made of different king of lipids, low, middle and high melting fractions, so phase transition occurs in a broad range of temperatures (-40 to $40°$ C). Besides composition, crystallization is affected by the cooling rate [2, 34]. Large crystals with heterogeneous spatial distribution develop under a slow cooling process, whereas smaller and more numerous particles with homogeneous distribution are promoted by a faster cooling. The newly formed crystals can protrude the globule surface and pierce the closest fat globules, favouring their partial coalescence and the formation of fat clusters [10, 33]. Milk samples subject to different industrial treatments can behave differently when stored at cold temperatures. Fucà et al. [8] found that weak treatments (i.e., pasteurisation and mild homogenisation) induced fat globule aggregation and fusion, when milk was stored at low temperature. Fat phase of milk samples subject to stronger treatments (i.e., UHT and intense homogenisation) was quite homogeneously distributed and no changes were detected at cold storage temperatures. These findings agree with Huppertz et al. [13], who found that creaming of milk fat globules on cold storage was more extensive in milk treated at 100-250 MPa than in milk treated at 400-600 MPa. The different behaviour of weakly- and strongly-treated milk is related to the complex IgM-lipoprotein (called cryoglobulin) that takes place along the cold storage [32]. Heat treatments at temperatures higher than $70°$ C and high pressure treatments provide for IgM denaturation and block its aptitude to bind milk fat globules [13].

4. Conclusion

Industrial treatments induce changes of fat globules microstructure. This chapter reviews recent findings on the effects of homogenisation, heat treatments and storage conditions on fat globules microstructure, focussing on MFGM. Several techniques have been introduced and their applications have been briefly described. Homogenisation is performed to prevent creaming. The main effect of mechanical stress is reduction of fat globules size. The exposed surface area increases and MFGM does not suffice to enclose the newly formed droplets. Missing MFGM material is supplied by caseins. Enzymes are released into the serum. Homogenisation pressure has a crucial impact on the entity and quality of the effects. Traditionally, milk sanitation is obtained by heating. High temperatures cause phospholipid loss replaced by adsorption of casein and whey proteins. Fat globules aggregation and fusion are also favoured by heating. Storage conditions affect stability of milk products. Melting and crystallisation of milk fat globules, originated from temperature variations, define the macroscopic properties of the final products. Depending on the treatment and on its entity, fat globules membrane goes through several structural changes, that have not been deeply understood. Novel technologies will allow to unveil new mechanisms and interactions among milk components.

References

[1] D. Bermúdez-Aguirre, R. Mawson, and G. V. Barbosa-Cánovas. Microstructure of fat globules in whole milk after thermosonication treatment. *Journal of Food Science*, 73(7), 2008.

[2] R. Campos, S. S. Narine, and A. G. Marangoni. Effect of cooling rate on the structure and mechanical properties of milk fat and lard. *Food Research International*, 35(10), 2002.

[3] M. E. Cano-Ruiz and R. L. Richter. Effect of homogenization pressure on the milk fat globule membrane proteins. *Journal of Dairy Science*, 80(11), 1997.

[4] M. Corredig and D. G. Dalgleish. Effect of different heat treatments on the strong binding interactions between whey proteins and milk fat globules in whole milk. *Journal of Dairy Research*, 63(3), 1996.

[5] D. G. Dalgleish and J. M. Banks. The formation of complexes between serum proteins and fat globules during heating of whole milk. *Milchwissenschaft*, 46, 1991.

[6] J. M. Evers, R. G. Haverkamp, S. E. Holroyd, G. B. Jameson, D. D. S. Mackenzie, and O. J McCarthy. Heterogeneity of milkfat globule membrane structure and composition as observed using fluorescence microscopy techniques. *International Dairy Journal*, 18(12), 2008.

[7] K. K. Fox, V. H. Holsinger, J. Caha, and M. J. Pallansch. Formation of a fat-protein complex in milk by homogenization. *Journal of Dairy Science*, 43(10), 1960.

[8] N. Fucà, C. Pasta, G. Impoco, M. Caccamo, and G. Licitra. Microstructural properties of milk fat globules. *International Dairy Journal*, in press, doi:10.1016/j.idairyj.2012.03.010.

[9] S. Gallier, D. Gragson, R. Jimenez-Flores, and D. Everett. Using confocal laser scanning microscopy to probe the milk fat globule membrane and associated proteins. *Journal of Agriculture Food Chemistry*, 58(7), 2010.

[10] R. W. Hartel. *Crystallization in Foods*. Aspen Publishers. Springer, 2001.

[11] A. V. Houlihan, P. A. Goddard, S. M. Nottingham, B. J. Kitchen, and C. J. Masters. Interactions between the bovine milk fat globule membrane and skim milk components on heating whole milk. *Journal of Dairy Research*, 59, 1992.

[12] Y. H. Hui. *Dairy science and technology handbook: Applications science, technology and engineering*. Wiley, New York, 1993.

[13] T. Huppertz, P. F. Fox, and A. L. Kelly. High pressure-induced changes in the creaming properties of bovine milk. *Innovative Food Science & Emerging Technologies*, 4(4), 2003.

[14] G. Impoco, N. Fucà, C. Pasta, M. Caccamo, and G. Licitra. Quantitative analysis of nanostructures' shape and distribution in micrographs using image analysis. *Computers and Electronics in Agriculture*, 84, 2012.

[15] R. H. Jackson and J. R. Brunner. Characteristics of protein fractions isolated from the fat/plasma interface of homogenized milk. *Journal of Dairy Science*, 43(7), 1960.

[16] T. W. Keenan, T.-W. Moon, and D. P. Dylewski. Lipid globules retain globule membrane material after homogenization. *Journal of Dairy Science*, 66(2), 1983.

[17] H. H. Y. Kim and R. Jimenez-Flores. Heat-induced interactions between the proteins of milk fat globule membrane and skim milk. *Journal of Dairy Science*, 78, 1995.

[18] S. J. Lee and J. W. Sherbon. Chemical changes in bovine milk fat globule membrane caused by heat treatment and homogenization of whole milk. *Journal of Dairy Research*, 69(4), 2002.

[19] A. V. McPherson and B. J. Kitchen. Reviews of the progress of dairy science: The bovine milk fat globule membrane - its formation, composition, structure and behaviour in milk and dairy products. *Journal of Dairy Research*, 50(01), 1983.

[20] M.-C. Michalski, R. Cariou, F. Michel, and C. Garnier. Native vs. damaged milk fat globules: Membrane properties affect the viscoelasticity of milk gels. *Journal of Dairy Science*, 85(10), 2002.

[21] M.-C. Michalski and C. Januel. Does homogenization affect the human health properties of cow's milk? *Trends in Food Science & Technology*, 17(8), 2006.

[22] M.-C. Michalski, M. Ollivon, V. Briard, N. Leconte, and C. Lopez. Native fat globules of different sizes selected from raw milk: thermal and structural behaviour. *Chemistry and Physics of Lipids*, 132(2), 2004.

[23] H. Mulder and P. Walstra. *The Milk Fat Globule: Emulsion Science as Applied to Milk Products and Comparable Foods*. Technical communication. Commonwealth Bureau of Dairy Science and Technology, 1974.

[24] L. Ong, R. R. Dagastine, S. E. Kentish, and S. L. Gras. The effect of milk processing on the microstructure of the milk fat globule and rennet induced gel observed using confocal laser scanning microscopy. *Journal of Food Science*, 75(3), 2010.

[25] R. Sharma, H. Singh, and M. W. Taylor. Composition and structure of fat globules surface layers in recombined milk. *Journal of Food Science*, 61(1), 1996.

[26] S. K. Sharma and D. G. Dalgleish. Interactions between milk serum proteins and synthetic fat globule membrane during heating of homogenized whole milk. *Journal of Agricultural and Food Chemistry*, 41(9), 1993.

[27] M. Thiebaud, E. Dumay, L. Picart, J. P. Guiraud, and J. C. Cheftel. High-pressure homogenisation of raw bovine milk. effects on fat globule size distribution and microbial inactivation. *International Dairy Journal*, 13(6), 2003.

[28] M. A. J. S. Van Boekel and T. Folkerts. Effect of heat treatment on the stability of milk fat globules. *Milchwissenschaft*, 46, 1991.

[29] S. Vargas Ruiz, R. Hass, and O. Reich. Optical monitoring of milk fat phase transition within homogenized fresh milk by Photon Density Wave spectroscopy. *International Dairy Journal*, 26(2), 2012.

[30] P. Walstra. Studies on milk fat dispersion. ii. the globulesize distribution of cow's milk. *Netherland Milk Dairy Journal*, 23, 1969.

[31] P. Walstra. Effect of homogenization on the fat globule size distribution in milk. *Netherland Milk Dairy Journal*, 29, 1975.

[32] P. Walstra. Physical chemistry of milk fat globules. In P. F. Fox, editor, *Advanced dairy chemistry. Lipids*, volume 2. Chapman & Hall, London, UK, 1994.

[33] P. Walstra. *Physical Chemistry of Food.* Food Science and Technology Series. Marcel Dekker, 2003.

[34] L. Wiking, V. De Graef, M. Rasmussen, and K. Dewettinck. Relations between crystallisation mechanisms and microstructure of milk fat. *International Dairy Journal*, 19(8), 2009.

[35] A. Ye, S. G. Anema, and H. Singh. High-pressureinduced interactions between milk fat globule membrane proteins and skim milk proteins in whole milk. *Journal of Dairy Science*, 87(12), 2004.

[36] A. Ye, H. Singh, M. W. Taylor, and S. Anema. Interactions of whey proteins with milk fat globule membrane proteins during heat treatment of whole milk. *Lait*, 84(3), 2004.

[37] A. Ye, H. Singh, M. W. Taylor, and S. G. Anema. Characterization of protein components of natural and heat-treated milk fat globule membranes. *International Dairy Journal*, 12(4), 2002.

INDEX

A

B

C

I

J

K

L

M

R

S

T